THE ECONOMICS OF FARM ANIMAL WELFARE

This book focuses on some of the latest developments in economic research that are relevant to animal welfare and related policy development. It consists of nine chapters; each has a specific focus related to various aspects of the economics of farm animal welfare. The authors explain in detail various topics, methods and approaches such as the evolution of animal welfare as a branch of animal science; animal welfare from an economic theory perspective; consumer demand and related quantitative methods such as willingness to pay, economics of production, supply side, regulations, policy and trade. Helpful to both natural scientists, such as animal welfare and animal behaviour scientists, as well as social scientists and agricultural economists, it serves as a reference for policy makers and policy analysts at both national and international level.

The authors

Bouda Vosough Ahmadi is Veterinarian and Animal Health Economist at the European Commission for the Control of Foot-and-Mouth Disease (EuFMD) of the Food and Agriculture Organization of the United Nations (FAO), Rome, Italy.

Dominic Moran is Professor of Agricultural and Resource Economics at Global Academy of Agriculture and Food Security, University of Edinburgh, UK.

Rick D'Eath is Reader in Animal Behaviour and Welfare at Scotland's Rural College (SRUC), Edinburgh, UK.

THE ECONOMICS OF FARM ANIMAL WELFARE

Theory, Evidence and Policy

Edited by

Bouda Vosough Ahmadi,

Dominic Moran

and

Rick D'Eath

CABI is a trading name of CAB International

CABI
Nosworthy Way
Wallingford
Oxfordshire OX10 8DE
UK
Tel: +44 (0)1491 832111
Fax: +44 (0)1491 833508
E-mail: info@cabi.org
Website: www.cabi.org

CABI
WeWork
One Lincoln St
24th Floor
Boston, MA 02111
USA
T: +1 (617)682-9015
E-mail: cabi-nao@cabi.org

© Bouda Vosough Ahmadi, Dominic Moran and Rick D'Eath 2020

British Library Cataloguing-in-Publication Data
A catalogue record for this book is available from the British Library, London, UK.

Library of Congress Cataloging-in-Publication Data
Names: Ahmadi, Bouda, editor.
Title: The economics of farm animal welfare : theory, evidence and policy / edited by Bouda Vosough Ahmadi, Dominic Moran, and Rick B. D'Eath. Description: Wallingford, Oxfordshire ; Boston : CABI, [2020] | Includes bibliographical references and index. | Summary: «The economic costs and benefits of farm animal production and sustainability versus improving climate change and animal welfare presents one of the most complex dilemmas in agriculture today. This book, by top global authors and experts, outlines the problem whilst making policy-relevant recommendations»-- Provided by publisher.
Identifiers: LCCN 2019056038 (print) | LCCN 2019056039 (ebook) | ISBN 9781786392312 (paperback) | ISBN 9781786392329 (ebook) | ISBN 9781786392336 (epub)
Subjects: LCSH: Food animals--Moral and ethical aspects. | Livestock--Economic aspects. | Livestock--Environmental aspects. | Animal welfare--Economic aspects. | Animal welfare--Government policy. | Sustainable agriculture.
Classification: LCC HV4757 .E26 2020 (print) | LCC HV4757 (ebook) | DDC 174/.9636--dc23
LC record available at https://lccn.loc.gov/2019056038
LC ebook record available at https://lccn.loc.gov/2019056039
References to internet websites (URLs) were accurate at the time of writing.

        978 1 78639 231 2 (paperback)
        978 1 78639 232 9 (ePDF)
        978 1 78639 233 6 (ePub)

Commissioning Editor: Alexandra Lainsbury
Editorial Assistant: Emma McCann
Production Editor: Marta Patino

Typeset by Exeter Premedia Services, Chennai, India
Printed and bound in the UK

Contents

Foreword

Animal welfare is at the heart of the legitimate expectations of society and also concerns about the future of livestock, both in developed and developing countries, in a global context where the demand for animal protein is increasing, particularly as a result of population growth, global economic development and the globalization of consumption patterns. The concept of animal welfare encompasses not only the physical health and wellbeing of the animal but also its psychological well-being and the ability to express the behaviours of the species. It relates to the relationships between humans and animals and all the controversies that surround them. The World Organization for Animal Health (OIE) seized upon the subject of animal welfare in 2002, particularly because of its close links with animal health. The Organization began by defining this notion, based on the five freedoms (absence of hunger and thirst; physical restraint; pain, injury and disease; freedom to express normal behaviour; protection from fear and distress). Since 2005, the OIE has developed international standards, based on scientific foundations, which serve as a reference at the global level. Animal welfare is a complex subject with various dimensions: scientific, philosophical, ethical, cultural, sociological, religious, political and economic.

This book, edited by Dr Vosough Ahmadi, Professor Moran and Dr D'Eath, has a particular focus on the economic and policy aspects of animal welfare. Respect for animal welfare should not always be experienced as a constraint on animal production; it can be an asset for livestock and for sustainable development. Issues around animal welfare concern breeding conditions and production systems, transport and slaughter, but also international trade, which leads to the growing integration of animal welfare requirements in the negotiation of bilateral trade agreements. In addition to ethical considerations, under many circumstances, improvement of animal welfare can lead to improvements in production and also the quality and safety of animals and animal products, resulting in 'win–wins'. There is a need for national, regional and global regulatory frameworks, policies and strategies. The concern to ensure the wellbeing of breeding animals is also fundamental. Veterinary services, considered a global public good, are both a key and a tool for improving animal welfare. These services must collaborate with all other services concerned, as animal welfare is closely linked to human health and wellbeing, respect for the environment and agricultural economics. In order to improve animal production alongside the rights and welfare of people, including farm workers and consumers, and the environment, the One Welfare holistic concept is developing, like that of One Health. A transdisciplinary approach is essential, with a permanent dialogue between all stakeholders. Animal welfare is fully in line with the agro-ecological transition approach and the United Nations Sustainable Development Goals.

This book will, I am sure, be a reference in the field for actors and decision makers concerned with animal breeding and the economy of animal production. I thank and congratulate the editors for their relevant initiative and wish the audience an excellent read!

Jean-Luc Angot
January 2020

Preface

Economists are preoccupied with efficiency of scarce resource allocation across a number of competing areas. This typically requires that costs and benefits of policy change are compared. When it comes to a public good like animal welfare, this becomes complicated since monetary costs are compared with monetary and non-monetary returns. The literature on animal welfare economics is light and very little has been done to address these important issues. On the other hand, from an animal welfare science point of view, assigning numerical value to the welfare impacts of policy changes in livestock systems is very challenging. This is because welfare is multifaceted (e.g. Five Freedoms) and the type and extent of welfare challenges (or improvements), and the number of animals affected by each of them is difficult to integrate into a single value of 'overall welfare' (or even a direction of change) at a farm or system level. This book attempts to bridge this gap to some extent by drawing on economic theory and economic research that is relevant to animal welfare and policy development. It consists of nine chapters that each has a specific focus related to aspects of economics of farm animal welfare.

Chapter 1 provides a background to the evolution of animal welfare as a branch of animal science and how animal welfare policy has developed, mostly driven by citizen pressure and activism. Chapter 2 discusses animal welfare from an economic theory perspective and provides an insight into the public good/private good debate as well as market failure. Chapter 3 considers consumer demand for animal welfare products and uses relevant quantitative methods such as willingness to pay for animal products with animal welfare attributes. Complementary to this, Chapter 4 presents and discusses peoples' preferences for animal welfare attributes. Chapter 5 covers the production and supply side with respect to the cost and benefits of improving animal welfare at farm level as well as in the food supply chain. It also addresses how the pursuit of economic efficiencies in production practices led to forms of cost-cutting that are at the root of much of the public unrest over animal welfare. Chapter 6 focuses on the economic impacts of welfare and trade-offs between welfare and environmental improvements by breeding and selection using an example from poultry production. Chapters 7 and 8 are devoted to regulations, policy and trade aspects of animal welfare in the EU and at the global level, respectively. Chapter 9 provides some thoughts on the expected future of animal welfare policies and discusses global objectives such as reducing environmental emissions and sustainable intensification with respect to animal welfare.

We hope that you find this collection of work an interesting and useful resource.

Acknowledgements

We would like to thank all those who contributed to this book and provided their valuable perspectives and results of their research. Secondly, we would like to thank our current and past employers and funders of our research for their support which allowed us to complete the book. We would also like to express our gratitude to many colleagues in various institutes who have contributed to our research in the field of economics of animal health and welfare during the last 20 years. We are especially thankful to Professor Alistair Stott for his critical review and valuable comments.

Finally, we are grateful to colleagues at CAB International, particularly Alexandra Lainsbury and Emma McCann, for their professional support during the book's preparation and production.

Contributors

Faical Akaichi is an agricultural economist currently working at Scotland's Rural College (Edinburgh, UK). Faical's research aims to further the understanding of producers' and consumers' choices and their determining factors, and how policy can best make use of this understanding to promote more sustainable and healthier food production and consumption.

Jean-Luc Angot is a veterinarian and General Inspector of Veterinary Public Health at the French Ministry for Agriculture. After having held various positions in ministerial administration, embassy, local veterinary services, inter-ministerial service and public agency, he became, in 2001, Deputy Director General of the OIE, and in 2009 French CVO and Delegate of France to the OIE. In 2015, he joined the High Council for Food and Agriculture, chairing the Prospective, Society and International Department. He chaired the FAO/EuFMD Commission from 2015 to 2019 and has been chairing the Codex Alimentarius Committee for General Principles since 2018. He is Head of the Body of the Inspectors of Veterinary Public Health and is the elected President of the Veterinary Academy of France for 2020.

Santiago Avendaño joined Aviagen in 2003 after graduating from Edinburgh University, UK, with a PhD in Quantitative Genetics and Genome Analysis. He is currently Global Director of Genetics for the Aviagen Group with responsibility for the development, evaluation and implementation of new technologies within Aviagen's breeding programmes. Prior to joining Aviagen, he was a researcher and technical advisor in beef cattle and sheep breeding in pasture-based production systems in Latin America.

Richard Bennett is Professor of Agricultural Economics at the University of Reading, UK. He has had a particular research and policy interest in relation to the economics and social science aspects of animal health, disease control and animal welfare over the last 35 years and has undertaken numerous research projects. He has served on a number of committees including the

UK Farm Animal Welfare Committee (for 10 years) and provides economic and policy advice to government and others.

Bettina Bock is Professor of Inclusive Rural Development at Wageningen University & Research as well as Professor of Population Decline and Quality of Life at Groningen University, both in The Netherlands. Her main research focus is on rural development in times of urbanization and rural–urban relations. She is also Editor-in-Chief of *Sociologia Ruralis*, one of the leading academic journals in the domain of rural social studies.

Donald M. Broom is Emeritus Professor of Animal Welfare at Cambridge University, Department of Veterinary Medicine, UK. His research concerns assessing animal welfare, animal cognition, sustainable farming and the scientific bases for morality and religion. He has published around 375 refereed papers and 12 books.

Beth Clark is a social scientist with an interest in stakeholder practices and perceptions of animal health and welfare. She is currently based in the Centre for Rural Economy at Newcastle University, UK. Her research focuses on how the public, farmers and farm advisors understand and view animal health and welfare, and how knowledge is gained and shared in relation to this. Her current research explores these topics in relation to endemic livestock disease in the UK.

Rick D'Eath is Reader in Animal Behaviour and Welfare at Scotland's Rural College (SRUC), UK. As an applied ethologist working on farm animals, his main research interests involve understanding how the farmed environment can modify and sometimes frustrate an animal's motivated behaviours, often leading to animal welfare problems. Rick primarily works on pigs and poultry, with a particular focus on questions around feeding and hunger, and interactions between animals that are negative for their welfare, including aggression, tail-biting and mounting.

David Harvey is Emeritus Professor of Agricultural Economics at Newcastle University, UK. He was President of the Agricultural Economics Society in 2004/5, is a Fellow of the Royal Agricultural Society of England and is the Editor of the *Journal of Agricultural Economics* (since 2005). He received an award for excellence from the Agricultural Economics Society in 2012 for 'outstanding contribution to public policy, industry and the profession'. His research interests focus on policy analysis and policy processes.

Carmen Hubbard is Senior Lecturer in Rural Economy at Newcastle University, UK. Trained as an agricultural economist, her research interests are in the economics and policy analysis of rural areas. She has extended her expertise to other countries such as Japan, Brazil, South Korea and Vietnam. In 2015 she was appointed to the Farm Animal Welfare Committee, the expert committee that advises the Department for Environment, Food and Rural Affairs (Defra) and the devolved administrations in Scotland and Wales on the welfare of farmed animals. In 2014, she was awarded a prestigious fellowship by the Japan Society for the Promotion of Science. She sits on the North East Farming and Rural Advisory Network steering group. She is Associate Editor of the *Journal of Agricultural Economics* and a member of the advisory board of EuroChoices.

Alfons Koerhuis is the Chief Technical Officer of Aviagen Group, responsible for the research and development of Aviagen chickens and turkeys. He has an MSc in Animal Breeding from Wageningen University and a PhD from Edinburgh University, UK.

Alistair Lawrence holds a chair in Animal Behaviour and Welfare at Scotland's Rural College and the Roslin Institute, University of Edinburgh, UK. While his main expertise is in animal behaviour applied to animal welfare, he has always had a motivation for interdisciplinary research, which has included work involving economics and other social science approaches. His current area of interest is in positive animal welfare to which he applies both conceptual and experimental approaches with the aim of clarifying what it means for animals to live 'good lives'.

Carolina Maciel is a Brazilian attorney who has been working as a researcher and consultant for more than 10 years on (farm) animal welfare policies, with a focus on the regulatory system of Brazil, the European Union and the recommendations and decisions of international organizations such as the World Trade Organization (WTO) and the Organization for Animal Health (OIE). Carolina holds a PhD in International Law and Policy from the University of Wageningen, The Netherlands, and an MSc in Political Sociology from the University of Santa Catarina, Brazil.

Dominic Moran is Professor of Agricultural and Resource Economics at the University of Edinburgh, UK. Previously, he worked for a period of 18 years at Scotland's Rural College (SRUC), prior to which he was a government economist and in private consulting. His research focuses on applying economics to environmental management and the development of interdisciplinary approaches to resource allocation problems in agriculture and global food security. Most recently, his work has focused on the challenges of reducing greenhouse gas emissions from agriculture and food supply chains, and the problem of antimicrobial use and resistance in agriculture. Dominic has worked in over 30 countries and has published more than 100 refereed journal papers. He has been in continuous receipt of funding from the EU, ESRC, NERC or BBSRC since 2000 for his research on climate change and agriculture and has supervised 20 PhD students.

Anne-Marie Neeteson is the Global Vice President of Welfare and Compliance at the Aviagen Group. She leads the International Poultry Council Environment and Sustainability Working Group, is a board member of the International Poultry Welfare Alliance and the US Round Table for Sustainable Poultry and Eggs and is active in various poultry working groups globally.

Jarkko K. Niemi is Research Professor of Economics of Sustainable Animal Production at the Natural Resources Institute Finland (Luke), specializing in the economics of animal health, welfare and production. He has particular interest in socio-economic aspects related to production diseases and animal welfare in pig and poultry production. He has worked on a wide range of areas such as economics of classical swine fever, African swine fever, bluetongue and foot-and-mouth disease in Finland, as well as economic analysis and optimization of pig production covering elements such as genetics, feeding, slaughter and replacement, technology choices, price, and production risks. He has also worked extensively in developing countries including Senegal, Kenya and Uganda.

Cesar Revoredo-Giha is Reader in Food Chain Economics, Senior Economist and Team Leader of Food Marketing Research at Scotland's Rural College (SRUC), Edinburgh, UK. His work focuses on agri-food value chains from primary producers to consumers, in developed and developing countries. In addition, he teaches topics on food security at the University of Edinburgh.

Alistair Stott retired as Head of Research at Scotland's Rural College (SRUC) in 2018. This followed nearly 40 years of research and education at the interface of agricultural science and agricultural economics. In particular, he pioneered the application of economics, systems modelling and management science to applied animal breeding, animal health and animal welfare.

Belinda Vigors is a social scientist currently working in the Animal Behaviour and Welfare Research Group at Scotland's Rural College in Edinburgh, UK. Her research interests include the use of qualitative research methods to examine how individuals make sense of and perceive farm animal welfare. In addition, she has a particular interest in the study of human decision making and how the perspectives of human behavioural science can contribute to better understanding how human behaviour impacts farm animal welfare.

Bouda Vosough Ahmadi is veterinarian and animal health and welfare economist currently working at the European Commission for the Control of Foot-and-Mouth Disease (EuFMD) of the Food and Agriculture Organization of the United Nations (FAO) in Rome. His main research focus is on agricultural policies and economics and policies of animal health and welfare. He currently leads socio-economic impact assessment studies of livestock diseases at EuFMD/FAO.

1

Farm Animal Welfare: Origins, and Interplay with Economics and Policy

ALISTAIR LAWRENCE,[1,2]* AND BELINDA VIGORS[1]

[1]*Animal Behaviour and Welfare, Animal and Veterinary Sciences, Scotland's Rural College (SRUC)*
[2]*The Roslin Institute, University of Edinburgh*

Summary

In this chapter we look at the origins of animal welfare as a societal concern and the interplay between the concept of animal welfare, economics and policy. We firstly propose adjustments to the 'standard view' of the development of animal welfare concerns (which we refer to as the Harrison-Brambell-FAWC (HBF) sequence). For example, we suggest that the role of science in setting animal welfare policy is a more complex process than is sometimes acknowledged. We discuss the application of economics to animal welfare including the analysis of the costs of animal welfare improvements to more recent work on trade-offs relating to animal welfare across the supply chain. Considering this range of uses of economics relating to animal welfare, we identify that the question of how to value animal welfare in economic terms remains unresolved. Lastly, we suggest that the period 1965–2008 may come to be regarded as a 'golden era' for the translation of animal welfare concerns into positive socio-political actions. We discuss a raft of issues which appear to have diminished the position of animal welfare in the policy 'pecking order'. However, societal concern over animal welfare will mean that government and others will need to be cautious of breaching 'red lines'. On a more positive note, the public profile that animal welfare enjoys will continue to provide the opportunity for policy and business innovations to improve animals' lives.

---

*Corresponding author: Alistair.Lawrence@sruc.ac.uk

© CAB International 2020. *The Economics of Farm Animal Welfare: Theory, Evidence and Policy* (eds B. Vosough Ahmadi *et al.*)

1.1   Introduction

We consider the origins of animal welfare as a societal concern and the inter-play between the concepts of animal welfare, economics and policy. Much of the material will be drawn from Europe and particularly the UK. While this may limit relevance to other geo-political areas, the UK and Europe have, arguably, the richest experience and history of animal welfare, and are there-fore most suited to exploring why animal welfare concerns have arisen and how they influence, and are influenced by, wider society.

1.2   The Origins of Animal Welfare: The Standard View

The standard view for the origin of animal welfare is that it originated in the mid-1960s in the UK, directly following the publication of Ruth Harrison's book *Animal Machines* (Harrison, 1964). The book illustrated that animal farming had moved significantly away from the public's perception of a 'rural idyll' to what thereafter became known as 'factory farming'. It gave rise, almost immediately, to misgivings among members of the British public about conditions in intensive farm animal production. It is rather remarkable that the book had such an immediate and profound effect, perhaps because of the public sensitivity to animal issues and perhaps because the 1960s had already seen increasing public alarm over other issues such as environmental pollution (Carson, 1962). One of the most significant and long-lasting im-pacts of Harrison's book was the forming, by the UK government, of the Brambell Committee, whose purpose it was to investigate and report on wel-fare conditions in British livestock farming. In 1965, the Committee issued its *Report of the Technical Committee to Enquire into the Welfare of Animals Kept under Intensive Livestock Husbandry Systems* (Brambell, 1965).

The Brambell Report is often seen as another seminal point in the de-velopment of animal welfare because it introduced a broader idea of what animal welfare should encompass. Whereas previous anti-cruelty legislation had focused on preventing what was seen as pointless or, as it was said, 'wanton' suffering without human benefits, this new development involved protecting animals against the adverse consequences of human activities even if the activities made food production more efficient. For example, al-though keeping sows confined using chains or crates, or housing slaughter pigs at very high stocking densities, could be seen as integral to the most efficient production of pork, these methods were still criticized for denying animals the fulfilment of their needs. The Brambell Report understood ani-mals' needs as something which, if they were not met, would cause suffer-ing. Thus the report insisted on a new and wider understanding of suffering, which went beyond persistent and significant pain to include the frustration of 'behavioural urges' in the form of discomfort, stress and, by inference, other negative mental states. This understanding of suffering made it pos-sible, for example, to criticize the confinement of sows, not on the basis that confinement causes pain but rather because confinement prevents animals from engaging in behaviours that they are highly motivated to perform. Using these ideas, the Brambell Report formulated the general requirements that farm animals should be free to 'stand up, lie down, turn around, groom themselves and stretch their limbs' (Brambell, 1965, p. 13).

The report also recommended the creation of a Farm Animal Welfare Advisory Committee (FAWAC), which was formed in 1967, to be superseded by the Farm Animal Welfare Council (FAWC) in 1979. It was FAWC which distilled the so-called Five Freedoms from the Brambell Report, formalizing these in a press release shortly afterwards (FAWC, 1979). Despite potential criticisms (e.g. McCulloch, 2013) and proposed alternatives (e.g. Mellor, 2016), the Five Freedoms have become the most widely used animal welfare framework globally (e.g. OIE, 2019), taking forward the broadening of animal welfare beyond prevention of cruelty. This includes the much-discussed freedom to 'express normal patterns of behaviour' (e.g. Bracke and Hopster, 2006; see below).

Many of the socio-political activities relating to animal welfare can be seen to follow from the 'Harrison-Brambell-FAWC' (HBF) sequence. In the UK, for example, Defra (2006) put the idea of animal needs into a legislative framework (House of Commons, 2006). The needs, as expressed in the act, for: a suitable environment (place to live); a suitable diet; to exhibit normal behaviour patterns; and to be: housed with, or apart from, other animals (if applicable); and protected from pain, injury, suffering and disease, are strongly influenced by the Five Freedoms.

Science had an important role in these developments. The Brambell Committee was clearly influenced by the emerging fields of animal behaviour and neuroscience. For example, in a footnote on page 10, the report states that the Committee was impressed by recent comments by Lord Brain (a neurologist) that he saw 'no reason for conceding mind to my fellow men and denying it to animals'. The committee also included W.H. Thorpe (then Director of the Animal Behaviour Department at the University of Cambridge) as an animal behaviour expert. Thorpe wrote an Appendix to the report entitled 'The Assessment of Pain and Distress in Animals', in which he developed the argument that animals are capable of suffering based on their physiology and behaviour, including those 'expressive movements' which are associated with deprivation and suffering. Dawkins (2016) argued that Thorpe set out an agenda for the future study of animal welfare that included the study of animals' subjective feelings.

In fact, much of the research that was to help fulfil Thorpe's agenda was not directly related to animal welfare but came about through developments in mainstream science. Most notable was the so-called 'cognitive revolution', which led to consciousness and awareness being widely accepted as suitable for scientific study across a range of disciplines and areas (e.g. Sperry, 1993). One consequence of this was for studies of animal behaviour to begin to explore animal cognition and awareness, refreshing the debate over animal mentality (e.g. Griffin, 2013). Thorpe's agenda was, however, also served by the more bespoke scientific area that we now refer to as animal welfare science. Animal welfare science is a loose amalgam of scientific disciplines ranging from the natural to the social sciences and including philosophy that since the late 1960s has focused specifically on addressing animal welfare issues (Lawrence, 2008a). The subject matter of animal welfare science followed closely Thorpe's agenda, particularly when scientists such as David Wood-Gush began to develop scientific approaches to study the 'animal's perspective', including studies of animal emotions such as 'frustration' (Duncan and Wood-Gush, 1971) and 'fear' (Hughes and Black, 1974). This early phase culminated in the publication of Marian Dawkins's seminal book *Animal Suffering: The Science of Animal Welfare* (Dawkins, 1980), in which she argued for animal preferences to be applied as a method for objectively assessing animals' motivational priorities and, by inference, their experiences.

Animal welfare science has continued to build on these early years through other innovative approaches to the study of the animals' experience (see Lawrence, 2016 for further details). For example, the development of judgement bias tests by Mike Mendl, Liz Paul and others is based on human psychology studies suggesting that underlying emotional states affect cognitive processing, for example with more depressed or anxious people judging ambiguous stimuli more negatively (e.g. Mathews and Mackintosh, 1998). The first application of 'cognitive bias' (as it is often referred to) tested rats for their response to ambiguous stimuli following exposure to different housing (Harding *et al.*, 2004) and has been followed by many studies applying cognitive bias across a range of species and contexts. Cognitive bias testing is supported by a theoretical framework aimed at understanding and interpreting research on animal emotions (e.g. Mendl *et al.*, 2010).

Another example of innovation in animal welfare science that builds on the HBF sequence is the development of qualitative behavioural assessment (QBA) by Françoise Wemelsfelder and colleagues, which directly fulfils Thorpe's aim to scientifically assess those 'expressive movements' which are associated with animal welfare. QBA arose from Wemelsfelder's position (part philosophical and part biological) that it can be legitimate to study animal behaviour from a qualitative perspective, and indeed that it may be essential to do so, in order to capture the subjective aspects relating to mental state that are of concern in animal welfare (Wemelsfelder, 2012). The result of this thinking led to the development of an approach to the recording of animal behaviour that focuses on the expressive quality of the behaviour as opposed to the quantitative descriptions of behaviour that are normally used in behavioural data collection (e.g. Wemelsfelder *et al.*, 2001). Similar to cognitive bias, testing QBA is now widely used in animal welfare science across a range of species from farm animals (e.g. Rutherford *et al.*, 2012) to elephants (Carlstead *et al.*, 2013).

In summary the standard view sees the publishing of *Animal Machines* as the essential primer to a sequence of events that was to define the development of animal welfare as a concept, as a public concern and as an area for scientific study.

1.3   Adjustments and Additions to the Standard View

In this section we want to suggest additions or adjustments to the standard view of animal welfare, partly to bring the standard view of animal welfare up to date with new evidence and also to better align the past development of animal welfare with the present.

1.3.1   The long view

So far we have described the HBF sequence as marking the transition from an era where animal welfare was seen, effectively, as the equivalent of animal cruelty to one where animal welfare took on a broader range of issues and concerns (e.g. Woods, 2012). However, this is possibly to downplay gradual changes in attitudes to animal mentality that had been happening since at least the 18th century (Lawrence, 2008b). For example, the Scottish philosopher David Hume wrote, in 1742, that 'animals undoubtedly feel … tho in a more imperfect manner than men' (Hume, 1987, revised edition). Perhaps the most famous early advocate of animals' moral status was Jeremy Bentham who wrote that 'the question is not, can they reason nor can they talk but can they suffer' (Bentham, 1789).

Radford (2001) suggests that early philosophers such as Hume and Bentham were influenced by emerging scientific evidence of the biological similarities between human and non-human animals, which called into question the anthropocentric view that dominated at the time. Thus, for at least 200 years, the idea that animals are sentient with some capacity to experience or feel has been evolving in philosophical circles, although how much more widely these views were to be found in society is harder to judge. It does seem plausible that this early belief in animal sentience was at least partly responsible for the first early steps to protect animals in law. As described by Radford (2001), the passing of the earliest animal protection legislation in the form of Martin's Act (1822) coincided with the writings of Bentham and others and the beginnings of a changing attitude towards animals' moral status.

1.3.2   Lack of clarity over what animal welfare is about

As we have seen, the standard view suggests that *Animal Machines* was responsible for the idea of animal welfare becoming more than a simple prevention of cruel acts. Woods (2012) has explored the historical evidence for this and presents a more nuanced analysis with some conclusions that resonate to the present day. For example:

> Although she [Harrison] is often retrospectively credited with being the first to articulate the concept of welfare, it should be noted that she was not the only critic of intensive farming, and that her book did not actually use the term 'welfare'. Rather, she expressed her concerns in the existing vocabulary

> of cruelty and suffering, while attempting simultaneously to redefine them.
> (Woods, 2012, p. 17)

Woods (2012, p. 18) debates why the UK government ministry responsible (MAFF) at the time used the term 'welfare' in the full title for the Brambell Report given that the term had previously been either little used or seen as a synonym for cruelty. One possibility suggested by Woods (2012, pp. 19–20) is that MAFF was seeking to ensure that they maintained responsibility for welfare over other departments (and particularly the Home Office), perhaps so that they could continue to best protect farmers against claims from welfare groups. Most importantly Woods (2012, p. 21) concludes that MAFF, in the period 1965–1971, never properly resolved its responsibilities for simultaneously representing farmers and farm animals. In particular, this meant that the tension between a farmer-facing, production-based view of welfare and the wider ethical perspective developed by Harrison remained unresolved. For Woods, this has contributed to the continuing contentious debate over farm animal welfare. Certainly, there is evidence that farmers see welfare more in terms of production efficiency and health than the broader welfare concept as developed through the HBF sequence (e.g. Skarstad *et al.*, 2007). Arguably, in the current era, with its focus on food security and even greater efficiency of production to 'feed the planet' (Thornton, 2010), this tension between different welfare perspectives is likely to be heightened. We will explore this issue further when we consider policy aspects of animal welfare.

1.3.3   The role of science is complicated

1.3.3.1   Animal sentience

The standard view tends to see science as facilitating and supporting the development of animal welfare (e.g. Radford, 2001). A closer look suggests that the role of science in animal welfare is more complicated. This is most evident with respect to the concept of animal sentience (e.g. Duncan, 2006). While much has been written in support of animal sentience and awareness (e.g. Griffin, 2013; Dawkins, 2006), there are also competing voices in science that deny either the existence of animal mentality or suggest that it is not a subject open to scientific investigation (e.g. Macphail, 1998).

While there are, today, relatively few dissenting voices that mammals and birds are sentient, the 'battle lines' become clearer as we move down the phylogenetic tree. One of the features of the animal welfare debate has been the wider inclusion of species previously 'ignored' with respect to their welfare, initially to include reptiles (e.g. Burghardt *et al.*, 1996) and fish (e.g. Mazeaud *et al.*, 1977), but also, more recently, invertebrates (e.g. Jones, 2013). The proposed extension of animal welfare to cover fish and invertebrates has been supported by scientific evidence for sentience in lower animals mainly in relation to pain (e.g. fish: Braithwaite and Huntingford, 2004; crustaceans: Elwood and Adams, 2015) but also, more widely, in cephalopods (e.g. Mather and Anderson, 2007) and, recently, insects (Barron and Klein, 2016).

However, the extension of sentience to fish and invertebrates again exposes the scientific complexity and controversy over animal sentience. In fish, for example, alternative perspectives, largely based on neuroanatomy (e.g. Key, 2015; Rose *et al.*, 2014), claim that it is more plausible (or parsimonious) that fish do not experience pain. Honey bees have been shown to display a negative cognitive bias when physically agitated (Bateson *et al.*, 2011), which opens up complex questions including whether honey bees have emotions or whether, instead, cognitive bias is a relatively 'primitive' response that has little connection with conscious experience (see Mendl *et al.*, 2011). Faced with such complex issues it is not unusual for scientific writings on animal sentience to conclude that we may never know for certain whether animals are sentient and if they are what the nature of their experiences are (e.g. Dawkins, 1993). For example, Bateson *et al.* (2011, p. 1072) write:

> Although our results do not allow us to make any claims about the presence of negative subjective feelings in honeybees, they call into question how we identify emotions in any nonhuman animal.

Despite these complexities, legislators have taken the step of enshrining animal sentience in law; first in EU law in the Treaty of Amsterdam (European Union, 1997) and then in the Treaty of Lisbon (European Union, 2007). Interestingly, in both cases, as far as we are aware, these treaties do not formally define which animals are covered. In the UK Animal Welfare Act (House of Commons, 2006), there is a clear definition of what is covered which makes direct reference to the role of scientific evidence:

> The Act will apply only to vertebrate animals, as these are currently the only demonstrably sentient animals. However, Section 1(3) makes provision for the appropriate national authority to extend the Act to cover invertebrates in the future if they are satisfied on the basis of scientific evidence that these too are capable of experiencing pain or suffering. Explanatory notes Section 1. (Defra, 2006)

The example of animal sentience rather clearly shows the complex interplay between scientific understanding of animal mentality and its further application in ethical debates relating to animal welfare and protection.

1.3.3.2   Science as a cultural modifier

Serpell (2004, p. S149) introduced the idea that science may have an indirect role in animal welfare development as a cultural modifier of our attitudes towards animals. He had in mind the role of science in providing information on the ecology and life histories of animals, which he suggested could help overcome indifference or dislike to specific animal species. He wrote about this just before the advent of internet-based video-sharing sites where the public have access to video material on animals' lives that was previously not available. A study of the popularity of YouTube videos of animal behaviour (Nelson and Fijn, 2013) found evidence that selected clips of animal play behaviour were watched by large audiences. Similarly, news reports of scientific studies of animal capacities such as 'intelligence' are now

to be found routinely on mainstream channels (e.g. Gill, 2018). However, as
Serpell (2004) points out, despite the potential value of the increased access
of the public to scientific information on animals, there is also the possibility
that this may be a double-edged sword, not least if the public become disen-
chanted by scientists' emphasis on detached 'objectivity'.

1.3.3.3   The debate over natural behaviour

The idea that animals should be able to express their natural behaviour and
live natural lives has been a major theme in the animal welfare literature and
also another major source of controversy. The origins are usually stated as
being the Brambell Report (e.g. Bracke and Hopster, 2006):

> In principle we disapprove of a degree of confinement of an animal which
> necessarily frustrates most of the major activities which make up its natural
> behaviour (Brambell, 1965, p. 13).

As a consequence, much of the early research on animal welfare was focused
on the importance of confined animals being able to perform natural behaviours, such as dust-bathing in hens (e.g. Vestergaard, 1982) and nest-building
in sows (Hansen and Curtis, 1980). One of the best-known pieces of animal
welfare research remains the 'Edinburgh Pig Park' study where researchers
studied the behaviour of commercial pigs released into a large enclosure with
the aim of studying how similar their behaviour was to wild pigs (Stolba
and Wood-Gush, 1989). The idea that domesticated animals should be able
to perform natural behaviours also resonates with what we understand to
be public perspectives on animal welfare where the importance of animals
living natural lives is a recurrent theme (e.g. Clark *et al.*, 2016; Lassen *et al.*,
2006).

Yet the focus on natural behaviour has also been controversial (e.g. Fraser
et al., 1997). From a scientific perspective, it has long been pointed out that natural living entails exposure to situations involving cold, hunger and predation,
which will result in unpleasant experiences and even death for animals (e.g.
Hughes and Duncan, 1988). The development of motivational approaches such
as 'preference testing' and 'consumer demand' to assess animals' priorities and
welfare was something of a reaction against the use of natural behaviour as the
central core of animal welfare research (Dawkins, 1980).

Fuller discussions of the role of natural behaviour in the animal welfare debate are available (e.g. Fraser *et al.*, 1997; Lassen *et al.*, 2006; Bracke and Hopster,
2006). We want to make an additional point here that appears to have been
missed in relation to the origins of the idea: while the Brambell Report refers
to the importance of natural behaviour (see above), it does also write at length
on the adaptations of domesticated animals to their current environments, e.g.:

> We accept the view that domesticated strains in general, and certain strains
> in particular, are much better adapted to caging than their wild ancestors
> would be, and doubtless suffer correspondingly less frustration… (Brambell,
> 1965, p. 19).

When it came to the distillation of the Five Freedoms by FAWC in 1979 (see above), the wording used was:

> Freedom to display most *normal* patterns of behaviour.[1]

The use of *normal* rather than *natural* in the Five Freedoms seems to be almost entirely overlooked (e.g. Bracke and Hopster, 2006), yet seems to us to be highly significant, particularly in the context of other writings in the rest of the Brambell Report. For us, *normal* places more emphasis on the animals' current 'state' than its previous (wild) 'state', allowing for the effects of processes such as artificial selection on the animals' responses to its environment. Thus, domesticated animals often retain the genetic basis for evolved behaviours (and we presume underlying motivational and learning processes) (e.g. Stolba and Wood-Gush, 1989). Yet at the same time, domestication has also led to adaptations of the animals' responses to the environment as evidenced by the widespread blunting of stress responses in domesticated species (e.g. Künzl and Sachser, 1999; Lepage *et al.*, 2000).

The use of normal behaviour in the Five Freedoms is also consistent with later writings. For example, the integrative model developed by Fraser *et al.* (1997) effectively makes the same point by talking about the animals' adaptations, however these have come about; for example, through evolution or domestication. Thus, we would argue in this respect that there is greater continuity than has previously been acknowledged between the Brambell Report, the Five Freedoms and more recent thinking on the relevance of domesticated animals' behavioural adaptations to their welfare.

1.3.3.4   Positive animal welfare

A relatively recent development in animal welfare has been the concept of positive animal welfare, which FAWC (2009) describes as animals having a 'life worth living' or even a 'good life'. There appears to be a number of different factors behind the rise in interest in positive animal welfare (see Lawrence *et al.*, 2018), including the wider acceptance that animals can experience emotions, including positive ones (e.g. Boissy *et al.*, 2007). Most relevant here are the suggestions that positive animal welfare is a necessary addition to the HBF sequence which, generally, if not wholly, focused on negation of harms such as fear, pain and stress. In this sense, positive animal welfare is similar to the arguments behind the growth in positive human psychology, which has similarly been accused of being overly focused on negative states such as depression (e.g. Myers and Diener, 1995). Much of the concern that the HBF sequence was unduly negative in its approach is directed at the phrasing of the Five Freedoms including by FAWC itself:

> One criticism of the Five Freedoms is their focus on poor welfare and suffering. This focus was undoubtedly appropriate at the time they were devised but the requirement to provide for an animal's needs in the new Animal Welfare Acts implies that good welfare should be an ambition too. (FAWC, 2009).

In fact, looking back at the Brambell Report and searching for text on positive behaviours, there are references to 'positive behaviours' such as animal play and curiosity. Some of these references to positive behaviours are made in the context of harms, e.g.:

> Both young and adult pigs show a good deal of play behaviour, and it has been stated that a play object such as a chain or rubber hose in each pen will occupy the attention of the group so well that it will minimise destructive activities such as tail-biting. (Brambell, 1965, p. 77)

However, some references give indication of a nascent interest in behaviours such as animal play, which appear to reflect positive emotional states:

> From the point of view of play, which is one of the best guides to the general intelligence level of a species, and which until not very long ago was regarded as peculiar to man … sheep rank very high probably higher than any other ungulate. (Brambell, 1965, p. 77)

The other exception to the claim that the Five Freedoms are wholly negative is of course the freedom to express normal behaviour which, as Yeates and Main (2008) point out, does introduce a more positive note while not specifically emphasizing the positive experiences that result from the performance of normal behaviour (see also Bracke and Hopster, 2006[2]). There is some scientific evidence to suggest that conditions such as environmental enrichment, which facilitates expression of normal behaviour, are associated with positive emotions. For example, Douglas *et al.* (2012) found that in young sows (gilts) an enriched environment was associated with an optimistic judgement bias, and that moving the gilts from an enriched to a barren environment induced a more negative reaction to barren housing than when experiencing only barren housing. However, in general, there is a need for further research to more fully understand the relationships between performance of normal behaviour and the brain systems underlying positive emotions and pleasure.

In summary, we have described the standard view of the development of animal welfare concerns as the Harrison-Brambell-FAWC (HBF) sequence. We have suggested adjustments and additions to the standard view including that development of animal welfare concerns pre-date the HBF sequence, that the HBF sequence did not result in a clear view of what animal welfare is about and that the role of science in the development of the animal welfare debate is rather more complicated than is often suggested. These perspectives on the development of animal welfare are the platform for the subsequent sections on the interaction between animal welfare, economics and policy.

1.4   Animal Welfare and Economics

Economics is a social science that studies the distribution of resources in the face of unlimited human demands (e.g. Lawrence and Stott, 2009). There are a number of ways in which economics has played a key role in the development of the animal welfare debate.

1.4.1   Counting the cost

As we have seen, the farm animal welfare debate from the mid-1960s has often focused on the lack of opportunities for farm animals to behave normally. This led to the development of 'alternative systems', which are designed to allow animals greater opportunities to express normal behaviour. For example, the Edinburgh Pig Park study (Stolba and Wood-Gush, 1989) was used to develop the Edinburgh Family Pen system (Stolba and Wood-Gush, 1984), which was specifically designed to allow pigs to express their normal behaviour (as defined by how pigs behaved in the Pig Park). Almost immediately this approach raised the two related issues of the practicalities of, and additional costs imposed by, such alternative systems. For example, with the Edinburgh Family Pen there were husbandry issues relating to piglet mortality, synchronization of pregnancies and general management of the system (Kerr et al., 1988), which, both practically and economically, combined to prevent commercial uptake of the system.

The use of economics to estimate the costs of alternative welfare-enhancing systems continues to the present. For example, one of the contentious aspects of modern pig production is the use of crates to house parturient (farrowing) sows. The widespread use of such systems emerged in the 1950s to ease the management of farrowing sows and, specifically, to provide protection to newborn piglets. The Edinburgh Pig Park study (Stolba and Wood-Gush, 1989) showed that part of the normal behavioural repertoire of domestic sows is to build a maternal nest (from available substrates such as branches) in which they give birth and nurse their young before emerging some days later. This nesting behaviour (which is also seen in wild pigs (D'Eath and Turner, 2009)) is triggered by the physiology that accompanies giving birth in the pig (Boulton et al., 1997). Consequently, even in the highly constraining farrowing crate, sows show increased activity and behaviour directed at the floor and the crate bars, which is interpreted as modified ('frustrated') nesting behaviour (Lawrence et al., 1994). This frustrated nesting is accompanied by increases in the stress hormone cortisol (Lawrence et al., 1994; Jarvis et al., 1997) and research has concluded that housing sows in farrowing crates, by the thwarting or frustrating of nesting behaviour, is stressful and may also interfere with subsequent maternal behaviour (e.g. Ahlström et al., 2002; Jarvis et al., 1999), which is very much in line with conclusions of the Brambell Report. There is therefore a sound body of evidence to support the development of alternative farrowing systems that allow sows to express nesting behaviour but at the same time provide protection to the piglet.

There has now been a number of attempts to evaluate the costs of production in different farrowing systems. For example, to compare the welfare and economic aspects of different farrowing systems, Baxter et al. (2012) developed a Welfare Design Index (WDI) to account for the biological needs of sows and piglets at different points (pre- during and post-farrowing). Different systems were then ranked for how well the biological needs of the individual animal were met and then compared to the physical and economic performance of the different systems. While acknowledging that the analysis was restricted by the lack of data, it did conclude that so-called 'designed pen systems' (e.g. PigSAFE), where there are defined areas for different sow activities and additional features to protect piglets, offered the best compromise between the competing demands of the sow, piglet and farmer. Subsequently, Guy *et al.* (2012) found a range of production costs from outdoor production (the cheapest) to PigSAFE (the most expensive option). However, the argument does not need to end with higher welfare systems being rejected because of their higher costs, as we will see below.

1.4.2   More on compromises and trade-offs

Economics has more to offer than analysis of costs and benefits. As mentioned earlier, the study of economics is essentially about decisions over how best to use scarce resources, or the trade-offs that exist when we are trying to achieve more than one outcome.

McInerney (1994) first described the concern over animal welfare (as initiated by the HBF sequence) traded off against the human-based interest of increasing farm animal productivity (see McInerney, 1994, Fig. 1, p. 16). McInerney suggested that initially taking animals under human care or management would increase their welfare (a 'win–win'), for example by reducing predation risk or reducing suffering from disease. It was where human activity increased productivity beyond certain levels that farm animals would begin to experience reduced welfare as described by Harrison and Brambell. Farm animal welfare could eventually become so reduced that it could be described as 'cruel'. McInerney's model is based on animal welfare as a good desired by humans; indeed that animal welfare only exists because humans are able to describe and care about it. An alternative view was suggested by Christensen *et al.* (2012) in which animal welfare is seen as an intrinsic value independent of human interests. This is reflected in the recent challenging of the anthropocentric approach (i.e. only human utility has value) of welfare economics (see Johansson-Stenman, 2018; McMullen, 2016). McMullen (2016), for instance, argues the demonstrated willingness of consumers to pay a price premium for higher welfare, and opinion polls detailing that some individuals believe animals deserve the same rights as humans indicate that people value animal welfare and wellbeing intrinsically, beyond its relevance to their consumption choices. However, the debate about how exactly animal welfare should be valued remains unresolved. None the less, the general principle of using economics to make animal welfare priorities transparent is a valuable approach, albeit one that is dependent upon an understanding of what matters to animals (Christensen *et al.*, 2012).

1.4.2.1   Farm level trade-offs

Trade-offs, potentially, exist at different points in the supply chain between farm and consumer, including at the farm level. Taking the example of alternative farrowing systems, Ahmadi *et al.* (2011) used linear programming to explore trade-offs between economic performance and managerial and animal welfare constraints using available data and expert opinion. The results suggested that if the target is to optimize piglet survival and the costs associated with ensuring piglet survival for the different systems, then crates provided the best net margin. However, non-crate systems were able to improve their net margin if account was taken of the extra welfare they provided for sow and piglets. This, of course, still begs the question of whether the value of the extra welfare will be recognized and paid for down the supply chain. We will return to this point.

Papers modelling the economics of alternative welfare systems often refer to the lack of available data (e.g. Baxter *et al.*, 2012; Ahmadi *et al.*, 2011). One aspect of the economics of alternative welfare systems that seems especially lacking in data is the biological interactions (trade-offs) between improving welfare and other biological functions, which could then have impacts on productivity and economics. There is something of a 'catch-22' about this, because without alternative systems being widely used by farmers it is not possible to accurately estimate reciprocal effects of improved welfare on other biological functions. In one of the few studies using a large sample of farms, Barnes *et al.* (2011) measured the production efficiency of dairy farms varying in levels of lameness. They found that farms with low levels of lameness were technically more efficient than farms with higher levels of lameness, while at the same time being inefficient in terms of their use of labour and stocking density. In other words, the use of greater inputs to reduce lameness (and increase welfare) was outweighed by the gain in milk yield on low-lameness farms. One explanation for this could be the greater use of grazing by low-lameness farms, which is known to help lame dairy cows recover (e.g. Hernandez-Mendo *et al.*, 2007), combined with lameness being known to reduce milk yield (e.g. Green *et al.*, 2002). Barnes *et al.* (2011) suggest that analysis of the farm-level efficiency of welfare traits should be done at the whole-farm level. Clearly, there is a need for more commercial-level data to explore the potential benefits (or disbenefits) of improved welfare on health and other functions.

1.4.2.2   Animal breeding: a special case of farm-level trade-offs

The use of artificial selection or animal breeding is an intrinsic aspect of modern farming (see Simm, 1998). Animal breeding is sometimes practised by the farmer but mostly by specialized breeding companies. Traditionally, animal breeding involves selecting animals for production traits (a trait being a measurable characteristic such as milk yield or growth rate). To 'improve' a trait is to change the trait in the anticipated direction (e.g. to increase milk yield or to reduce lameness). Often, animal breeding will select on more than one trait at a time, which has given rise to multi-trait selection (see Lawrence *et al.*, 2004 for a more detailed description). In brief, multi-trait selection uses the individuals' breeding value (BV) for each trait multiplied by the trait's economic value. Individuals are then ranked on the sum of these values. While the selection is aimed at measurable traits, other traits may also be affected because they are genetically linked or correlated to one or more of the selected traits. As pointed out by Rauw *et al.* (1998), this can have undesirable consequences in terms of reducing the overall fitness of the animal. For example, a narrow selection for milk yield in dairy cattle can lead to a genetic propensity for increased lameness and mastitis (Sandoe *et al.*, 1999). Thus, animal breeding is another form of farm-level trade-off where farmers (or breeding companies) make choices about the traits they want to emphasize in their animals, given that selecting for improvement in one trait will usually mean less improvement or even loss of function in others. The trade-offs involved in animal breeding are more straightforward for conventional 'market' traits (e.g. growth) but become more complex when less conventional 'non-market' traits are considered, for example relating to animal welfare (Sandoe *et al.*, 1999) or the environment (Wall *et al.*, 2010). One reason that trade-offs are more complex for traits such as animal welfare is that the economic values are harder to estimate. Taking a trait such as lameness in dairy cattle, while the direct economic costs of lameness are built into multi-trait selection, there is no estimate for the indirect cost to the cow in terms of pain and other relevant welfare experiences (e.g. Lawrence *et al.*, 2004) and no way of knowing if the direct economic costs are also equivalent to the indirect costs. In general, animal breeding again raises the question of how to 'value' animal welfare.

1.4.2.3   Trade-offs beyond the farm level

Trade-offs clearly also exist beyond the farm level, but there has been relatively little work to evaluate these trade-offs with respect to animal welfare. Toma *et al.* (2008) showed that improving piglet survival (arguably a welfare improvement) could have wider beneficial effects on trade and the environment. But generally, there has been more focus on the potential negative effects of higher welfare standards in the EU, either as an impediment to trade (e.g. Van Horne and Achterbosch, 2008) or as a driver for change in other countries such as the USA (Mench *et al.*, 2011). Finally, of course, consumers are likely to be making trade-offs in terms of their consumption patterns, most notably in the well-described gap between attitudes to animal welfare and buying of animal products (Vigors, 2018). Various studies suggest that although consumers may express concern for animal welfare in principle, when it comes to the point of purchase, consumers will trade off animal welfare for other attributes, particularly price (e.g. Akaichi and Revoredo-Giha, 2016).

1.4.3   More on valuing animal welfare

The question of how to value animal welfare is clearly a central issue in relation to applying economics to animal welfare. As we discussed earlier, one economic view of animal welfare is that it is a subset of our own perceptions and only indirectly to do with animals (McInerney, 1994). This view was based on the lack of evidence that animals are able to make rational choices or that we cannot assess their welfare in other robust ways (see McInerney, 1994, p. 13). As previously noted, animal welfare science has developed a number of innovative approaches to assess animal welfare and today it is much harder to argue against the idea that animals have both good and bad experiences. In other words, animal welfare exists independently of humans, for example, as in the case of welfare of wild animals outside of our control or management (Sainsbury *et al.*, 1995). Thus the development of animal welfare science has moved the argument closer to the position suggested by Christensen *et al.* (2012), that animal welfare is an intrinsic value, with the consequence that the level of animal welfare that a society aims for should be based on more fundamental principles (e.g. scientific understanding) and not only determined by the way individual consumers trade off animal welfare with other desirable goods. Consequently, there is a developing argument for the inclusion of a measurement for animal welfare's intrinsic value in economic models and a rethinking of its anthropocentric approach (see Johansson-Stenman, 2018).

A related argument has been made by Buller and Roe (2012) that the 'selling' of animal welfare (e.g. through assurance schemes and labelling) is a commodification of animal welfare that focuses on those elements which can be commodified, and may ignore equally important but less saleable aspects. This has some similarity to the problem we discussed earlier in relation to animal breeding where economic values are only applied to marketable traits (i.e. where welfare impacts on productivity) and not non-market traits such as the animals' experiences. There have been various suggestions about how to estimate economic values for non-market welfare attributes. For example, Lawrence *et al.* (2004) pointed to the use of approaches such as 'contingent valuation' and 'willingness to pay' to estimate non-market ethical values (e.g. Bennett and Larson, 1996). A recent meta-analysis of willingness to pay (WTP) studies applied to farm animal welfare (Lagerkvist and Hess, 2010) does suggest a general WTP for farm animal welfare, albeit with a number of complexities that warrant further research in this area. Notably, Frey and Pirscher (2018) recently put the intrinsic value of animal welfare to the test and concluded that moral values, specifically altruism, reduced apathy and environmental concern, do influence and increase consumer WTP for improved animal welfare.

1.4.4 Changing human behaviour

The evidence of WTP for improvements to farm animal welfare suggests that incentives could be used to transfer benefits within the supply chain to encourage farmers to uptake alternative systems.

Christensen *et al.* (2012) discuss the use of incentives to ensure more even transfer of the benefits and costs of improving animal welfare across the supply chain. In some cases, the incentives to change farmers' behaviour should reside at the farm level where improvements to welfare will bring with them obvious benefits; examples include genetic selection for health traits, improving neonatal survival and genetic selection for certain types of animal temperament (see Lawrence and Stott, 2009). However, even where there are clear economic benefits to farmers, there may be constraints preventing uptake of improved welfare practices. One example of this is lameness in dairy cattle, which, as we have discussed, is well characterized as leading to loss of productivity (e.g. Green *et al.*, 2002), and yet the levels of lameness on farms have remained constant since the 1990s (Leach *et al.*, 2010). In relation to genetic selection for animal temperament (e.g. increased docility), there may be ethical constraints to be resolved (e.g. D'Eath *et al.*, 2010).

In other cases, the incentives to change system or husbandry practice may not reside at farm level. Earlier we used the example of alternative farrowing systems to illustrate the application of economics to farm-level trade-offs. One of the conclusions of work in that area (e.g. Guy *et al.*, 2012) is that alternative indoor designed pen systems such as PigSAFE are more expensive to run, which suggests that a price premium may be needed to encourage uptake (Cain *et al.*, 2013). The obvious route for this would be through large retailers choosing to pay farmers to produce pork from farms that use alternative farrowing systems (Christensen *et al.*, 2012). The limitations to this market-led solution to incentivizing uptake of alternative systems is, as we have already discussed, that it will be variable with respect to the welfare issue and dependent upon retailers being convinced that there is some market advantage in their support of the alternative systems.

The wider discipline of economics has undergone dramatic changes since the late 1970s, where the limits of normative economics to predict human behaviour prompted a merging with psychology to create the sub-field of behavioural economics (Mathis and Steffen, 2015). Within animal welfare economics, this is reflected in the increasing recognition that understanding and changing behaviour requires knowledge of the underlying motivations of stakeholders (Christensen *et al.*, 2019). Behavioural economics suggests that, rather than being motivated by maximizing benefits and minimizing costs, an individual's decisions are driven by, for example: heuristics; how information is presented; past experiences; the context of the decision; the behaviour of others; and the cognitive reference points they draw from (Gigerenzer and Gaissmaier, 2011; Ratner *et al.*, 2008; Thaler and Sunstein, 2008). The challenge is to unpick such motivations and influences in the context of animal welfare and to determine how best to harness them to improve animal welfare-related behaviours.

Within the consumer literature, some research suggests that increasing the salience of product attributes desired by consumers can influence their choice. Akaichi *et al.* (2019), for example, found that demand for organic meat increased when its animal welfare and nutritional benefits were emphasized, while Carlsson *et al.* (2007) found that Swedish consumers, due to concerns with live transportation, would pay a premium for cattle slaughtered by a mobile abattoir. In addition to emphasizing specific food attributes, there is evidence that alternately describing the same meat product attribute positively or negatively changes consumer WTP (Fox *et al.*, 2002). Such findings reflect insights in behavioural economics that human choice is influenced by how information is conveyed (e.g. positively or negatively) and what attributes such descriptions make salient (e.g. Böhm and Theelen, 2016; Tversky and Kahneman, 1981). As such, how information is conveyed, it is argued, influences behaviour change (Thaler and Sunstein, 2008).

Farmer welfare-related behaviour can also be motivated by nuanced, intrinsic factors, beyond the extrinsic motivations (i.e. incentive) of enhancing profits. For instance, de Lauwere *et al.* (2012) found that farmers who had changed to group housing for sows, in part, felt that important stakeholders had expected them to, while those who had not changed were less likely to think this. Fear of a negative evaluation from others can also motivate farmers to change behaviours (Hansson and Johan Lagerkvist, 2012), particularly when there is potential for clear utility losses such as reputational damage in the eyes of consumers (Belay, 2018) or the loss in social status as a 'good farmer' (Te Velde *et al.*, 2002). Christensen *et al.* (2019) thus note that having a 'social licence' to farm is increasingly recognized to have utility value, leading to the growing inclusion of reputational functions in economic analysis.

As is increasingly notable within the welfare economics literature, human behaviour is complex and often context-dependent. This has implications for how to motivate behaviour change and what interventions to use. Extrinsic incentives, for instance, may not be a useful tool in contexts where individuals are motivated by information salience and the expectations of others. An alternative to the use of incentives to change behaviour within the supply chain is the application of 'nudging', a set of behavioural change approaches derived from behavioural economics. Vigors (2018) recently suggested that nudging, which has been used successfully to better align individuals' intentions and action in other similar areas, could be applied to animal welfare. The attraction of 'nudging' is that it accepts the complexities of human choice behaviour and aims to work with these rather than against them.

In summary, economics has been applied in a number of ways to the animal welfare debate. From an initial focus on the costs of alternative welfare systems, it has been used to explore trade-offs including those involving consumer behaviour with respect to purchase of animal products. Among outstanding issues requiring further research are questions relating to the valuation of animal welfare and how to close the attitude–behaviour gap in order to more evenly distribute benefits and costs of improved animal welfare through the supply chain.

1.5   Animal Welfare and Policy

As we have discussed, animal welfare in economic terms is a 'non-market' or 'public' good, which means that it is incompletely reflected in the market, and hence it is likely that there will be 'market failure' where the market fails to provide as much animal welfare as society expects or demands (e.g.

Bennett, 1995). The non-market or ethical aspects of animal welfare therefore demand attention by government on how best to deal with animal welfare as a public good. FAWC (2008) has produced a summary (from a UK perspective) of different policy options available to government, specifically with respect to protecting and improving animal welfare. Here we provide a retrospective and prospective review of changes to animal welfare policy, again through the lens of the UK and European experience.

1.5.1   The 19th and 20th centuries

We have already seen that government from the 19th century was involved in attempting to resolve society's concerns over animal welfare. Although the first animal protection legislation in the UK (Martin's Act 1822) was promoted by Richard Martin as a private member, within the next 50 years the UK government was to pass further acts, in 1835 and 1849, to make farm animal protection more effective (Radford, 2001). Later in the 19th century, the UK government was also drawn into the argument over the use of vivisection, and in response passed the Cruelty to Animals Act (1876) to limit the uses of vivisection (Defra, 2019a).

Thus the initial response of the UK government was the use of legislation in response to concern over animal cruelty and welfare, and this pattern continued through the 20th century with the major proviso that from the early 1970s UK animal welfare legislation has largely been taken from EU law. There have been occasional examples of the UK taking the legislative 'lead' over animal welfare, most famously with the banning of the use of stalls for pregnant sows in 1999 (Brooks, 2003), which was some years in advance of the EU ban on this housing system. However, largely as with other EU countries, the UK has followed the EU lead, including the enshrining of animal sentience in law (see earlier).

Other notable policy developments in the 20th century were, of course, triggered by the publishing of *Animal Machines* (Harrison, 1964). We have already described how this led, in the UK, to the setting up of the Brambell Committee and, later, the Farm Animal Welfare Council. Woods's (2012) historical account of how different UK government departments dealt with the developing debate over animal welfare between 1965 and 1971 is an example of how 'messy' the translation of a public interest issue such as animal welfare into policy can be. In particular, as we have seen, the conflicted position of MAFF resulted in the unresolved issue of whether animal welfare is about the animal's physical state and productivity or the ethics of how we manage sentient animals that are capable of pleasant and unpleasant experiences (Woods, 2012).

1.5.2   2000–2008

In 2001, the UK suffered a widespread outbreak of foot-and-mouth disease (FMD). The FMD outbreak was hugely damaging economically and socially

(Woods, 2004), and as a result it ushered in a change of government policy that aimed to change the balance from government to shared responsibilities for animal health and welfare (The GB Health and Welfare Strategy, Defra, 2004). Re-reading of this strategy provides justification for the perspective that the UK government at the time remained conflicted in its idea of what animal welfare is about. Although animal welfare is introduced in the strategy in terms of the Five Freedoms, much of the text refers to animal health and welfare purely in terms of physical fitness and performance and the importance of animal welfare with respect to disease risk, e.g.:

> Animals that are cared for appropriately and in accordance with acceptable welfare standards are more likely to be healthy, and less likely to contract or spread disease (Defra, 2004, p. 9).

> Fit and healthy animals which are appropriately cared for are likely to be higher yielding or remain productive over a longer period of time (Defra, 2004, p. 22).

None the less, despite its rather superficial treatment of animal welfare, the GB Health and Welfare Strategy is an important document because it was an early signal of a wider change in government policy with respect to animal welfare, moving away from the almost total reliance on legislation to a range of other policy options (see also FAWC, 2008). This policy change is most clearly articulated in the EU's own Strategy for Protection and Welfare of Animals 2012–2015 (European Commission, 2012), which clearly questions the use of a prescriptive legislative approach to improving animal welfare:

> [I]t has become increasingly clear that simply applying the same sector specific rules to animal welfare does not always yield the desired results. Problems of compliance to sector specific rules point the need to reflect on whether a 'one size fits all' approach can lead to better welfare outcomes across the Union. (p. 4)

As an alternative strategic direction, the EU 2012–2015 Strategy refers to development of a revised EU legislative framework based on a holistic approach and specifically introducing science-based welfare indicators based on outcomes as opposed to the use of inputs (European Commission, 2012, p. 6). In other words, rather than prescribe the physical conditions in which animals are housed to ensure welfare, future legislation could allow the use of scientifically based (biological) welfare indicators to assess animal welfare. We can see a common link between this proposal and the HBF sequence, which led to the development of animal welfare science, with its focus on the development of scientific welfare indicators. At the same time, such a policy shift could be seen as giving farmers greater freedoms in how they keep animals, provided that animal-based welfare indicators are within certain bounds. In time, there could also be concerns that, through direct or indirect effects of breeding, animals become adapted to what might at the outset be regarded as poorer welfare conditions (see also D'Eath et al., 2010). The linkage of this approach to concerns over competitiveness (see p. 7 of the Strategy) suggests again something of a conflict of interest in terms of policy representing the farming industry or representing the animals.

The other strategic direction proposed in the Strategy is a set of specific actions that include providing 'consumers and the public with appropriate information' (see also Section 1.4.4 on 'Changing human behaviour'). The concept of using market-led approaches to improve animal welfare can be linked to the creation of farm-assurance schemes in the 1980s. Although such schemes were originally devised for other purposes (e.g. to promote product provenance), by the early 2000s, farm assurance schemes had already begun to take on animal welfare assessment (FAWC, 2005). Despite the apparent simplicity of using farm assurance to assure consumers over farm animal welfare, there are in fact many technical, economic and ethical issues involved. The FAWC report on farm assurance and animal welfare provides a review of science-based methods for assessing welfare on farms and specifically the development of animal-based measures (FAWC, 2005, Appendix A, p. 51). In summary, the main constraints relate to development of welfare measures which are reliable and valid and can be applied effectively in the brief period that inspectors are on farms. It has proved difficult to find measures that meet these demanding criteria. A related issue is the question of how to summarize and integrate information on welfare in ways that can be quickly assimilated by consumers, given the multi-dimensional nature of animal welfare (FAWC, 2005; Blokhuis *et al.*, 2010). In wider economic and ethical terms, there is also the risk with market-led approaches of creating a multi-tier system where some animals are managed to minimum welfare standards while others live a 'good life' (e.g. Edgar *et al.*, 2013).

In summary, we can see the period between the publishing of *Animal Machines*, in 1964, to 2008 as representing a 'golden era' for animal welfare supported by a positive policy environment. We base this assessment on the substantial government activity in the forming of expert committees and passing of substantial amounts of legislation, which included the placing of animal sentience into legislative frameworks. This was followed by non-legislative government-level strategies and action plans at national and international levels. Throughout this period there was also substantial research funding for animal welfare, much of it aimed at the development of scientifically valid methods for assessing animal welfare.

1.5.3   2008 to the present and beyond

This last period starts with the financial crisis of 2008, where there is evidence that animal welfare was moved down to a lower level of policy priority, at least in the UK and possibly at EU level as well. Evidence for this includes, at the UK level, a substantial reduction in research funding for animal welfare. At the time of the financial crisis, the Department for Environment, Food and Rural Affairs (Defra) had a budget of over

£3 million for farm animal welfare, a sum that had remained stable since 2000/2001 (FAWC, 2008). Annual calls would go out on policy-relevant issues such as the assessment of the welfare impacts of transport. Current levels of research funding appear to be substantially lower, at less than £1 million, and it has been a number of years since there have been the annual welfare research calls. The Defra website (Defra, 2019b) reveals that few animal welfare projects are currently being funded; indeed the Defra web pages devoted to animal welfare have not been updated since December 2017 (Defra, 2017). In addition, it would appear that the number of UK government personnel working on animal welfare has dropped substantially, and there has also been a general loss of senior veterinary expertise in animal welfare. FAWC has been moved to a different arrangement with respect to government now being classed as a Committee as opposed to a Council, which may have some implications for its autonomy. At the EU level, there has been no follow up to the 2012–2015 animal welfare strategy (European Commission, 2012), although there has been other activity including setting up of an animal welfare platform (Ribeiro, 2017) and an EU Reference Centre for Animal Welfare (Binns, 2018).

The proposition that animal welfare has declined as a policy priority is based on discussions with policy personnel and a small amount of available data, and must therefore be treated with caution. If there is substance to the proposition, then possible reasons could be that this is mainly a UK issue brought on, in part, by the financial crisis and possibly now exacerbated by the complications surrounding Brexit. More widely, it could reflect the growth of other competing policy concerns, continuing concerns over the impact of farm animal welfare on competitiveness, the belief that UK (and possibly EU) standards are now sufficient, a growing reluctance to legislate on such issues and effective lobbying by agri-businesses. We would propose that a full review of animal welfare policy interactions is warranted to better understand these trends.

Given our uncertainty over the current relationship between animal welfare and policy, predicting the future is fraught with difficulties. However, it seems to us that animal welfare will struggle to retain or perhaps regain the rather privileged policy position it has previously occupied in the UK and the EU into the future, for all the reasons we have listed above. Despite this, the available evidence suggests that the public will remain highly sensitive to animal welfare (at least in the UK) and policy makers will need to operate with care. For example, in the parliamentary process of voting for EU withdrawal, the UK government did not support the transfer of Article 13 of the Lisbon Treaty, recognizing animals as sentient beings, into UK legislation (Ares, 2018). There are a number of interpretations of this move, but it raised considerable adverse public response which culminated in the government appearing to step back and sponsor the Animal Welfare (Sentencing and Recognition of Sentience) Draft Bill (Ares, 2018).

1.6   Conclusions

The aim of this chapter was to provide an overview of the development of animal welfare and its interactions with economics and policy. In describing the development of animal welfare we wanted to illustrate the value of going 'beneath the surface' of the standard view (which we referred to as the Harrison-Brambell-FAWC (HBF) sequence). For example, we discussed the complexities of the relationship between scientific knowledge of animal welfare and policy. On the one hand, while science continues to struggle with the subjective nature of animal wellbeing, the concept of animal sentience has been incorporated into law in the EU and the UK, effectively avoiding the scientific arguments over the nature or even existence of animal mentality. We also used a recent historical analysis by Woods (2012) to illustrate how the translation of animal welfare into government policy in the UK in the period 1965–1971 was affected by interdepartmental rivalries and also by conflicted responsibilities to farm animals and to farmers. As a consequence, it can be argued, the UK failed to emerge from this early phase with a clear idea in policy terms of what animal welfare was about, potentially exacerbating the contentious nature of the animal welfare debate.

We then explored the relationship between animal welfare and economics, from the use of economics to analyse the costs of animal welfare improvements to more recent work on trade-offs relating to animal welfare across the supply chain. Across this range of uses of economics relating to animal welfare, we identified that the question of how to value animal welfare remains unresolved. We saw this specifically in relation to the application of economics to animal breeding, but also more generally in the question of whether animal welfare has intrinsic value separate from human preferences. The nuances of human preferences and behaviours, as described by behavioural economics, were also noted to influence how individuals respond to animal welfare within the supply chain. The relevance of such innate features of human decision making to changing human behaviour in relation to animal welfare was further discussed.

Lastly, we described a timeline for policy developments in animal welfare from a UK perspective. We posited that the period 1965–2008 may come to be regarded as a 'golden era' for translation of animal welfare concerns into positive socio-political actions. We discussed a raft of issues which appear to have diminished the position of animal welfare in the UK policy 'pecking order'. We further suggested that this downward shift in the policy priority given to animal welfare may go beyond the UK, and that it is unlikely that animal welfare will regain its previous policy position. To counter this, however, it seems that in the UK at least there will continue to be an underlying and widespread sensitivity to animals with a matched concern for their welfare. This societal concern will mean that government and others will need to be cautious of breaching 'red lines' and, on a more positive note, still provide the opportunity for policy and business innovations designed to improve animals' lives.

Notes

[1] The current form of words used to the present day by FAWC (and repeated wherever the Five Freedoms have been used by other bodies such as the OIE) is the 'Freedom to express normal behaviour'.

[2] Bracke and Hopster (2006) assume that the Five Freedoms' use of normal behaviour is equivalent to natural behaviour. We do not agree, as we explained earlier.

Acknowledgements

We acknowledge funding support for this work from the Scottish Government's Rural and Environment Science and Analytical Services Division (RESAS) and the BBSRC Strategic funding to the Roslin Institute. We would also like to acknowledge helpful inputs from Peter Sandøe and Henry Lawrence.

References

Ahlström, S., Jarvis, S. and Lawrence, A.B. (2002) Savaging gilts are more restless and more responsive to piglets during the expulsive phase of parturition. *Applied Animal Behaviour Science* 76(1), 83–91. DOI: 10.1016/S0168-1591(01)00207-6.

Ahmadi, B.V., Stott, A., Baxter, E., Lawrence, A. and Edwards, S. (2011) Animal welfare and economic optimisation of farrowing systems. *Animal Welfare* 20, 57–67.

Akaichi, F. and Revoredo-Giha, C. (2016) Consumers demand for products with animal welfare attributes: evidence from homescan data for Scotland. *British Food Journal* 118, 1682–1711.

Akaichi, F., Glenk, K. and Revoredo-Giha, C. (2019) Could animal welfare claims and nutritional information boost the demand for organic meat? Evidence from non-hypothetical experimental auctions. *Journal of Cleaner Production* 207, 961–970. DOI: 10.1016/j.jclepro.2018.10.064.

Ares, E. (2018) *Animal Sentience and Brexit.* (Briefing Paper No. 8155). House of Commons Library, London.

Barnes, A.P., Rutherford, K.M.D., Langford, F.M. and Haskell, M.J. (2011) The effect of lameness prevalence on technical efficiency at the dairy farm level: an adjusted data envelopment analysis approach. *Journal of Dairy Science* 94(11), 5449–5457. DOI: 10.3168/jds.2011-4262.

Barron, A.B. and Klein, C. (2016) What insects can tell us about the origins of consciousness. *Proceedings of the National Academy of Sciences* 113(18), 4900–4908. DOI: 10.1073/pnas.1520084113.

Bateson, M., Desire, S., Gartside, S.E. and Wright, G.A. (2011) Agitated honeybees exhibit pessimistic cognitive biases. *Current Biology* 21(12), 1070–1073. DOI: 10.1016/j.cub.2011.05.017.

Baxter, E.M., Lawrence, A.B. and Edwards, S.A. (2012) Alternative farrowing accommodation: welfare and economic aspects of existing farrowing and lactation systems for pigs. *Animal* 6(1), 96–117. DOI: 10.1017/S1751731111001224.

Belay, D. (2018) Economics of information and incentives in regulation of market failure: information disclosure, impact evaluation, market design, antibiotics and commodity markets. PhD thesis. The University of Copenhagen.

Bennett, R. (1995) The value of farm animal welfare. *Journal of Agricultural Economics* 46(1), 46–60. DOI: 10.1111/j.1477-9552.1995.tb00751.x.

Bennett, R. and Larson, D. (1996) Contingent valuation of the perceived benefits of farm animal welfare legislation: an exploratory survey. *Journal of Agricultural Economics* 47(1–4), 224–235. DOI: 10.1111/j.1477-9552.1996.tb00686.x.

Bentham, J. (1789) *Introduction to the Principles of Morals and Legislation*. Clarendon Press, Oxford, UK.

Binns, J. (2018) EU reference centre for animal welfare, food safety – European Commission. Available at: https://ec.europa.eu/food/animals/welfare/eu-ref-centre_en (accessed 1 April 2019).

Blokhuis, H.J., Veissier, I., Miele, M. and Jones, B. (2010) The welfare Quality® project and beyond: Safeguarding farm animal well-being. *Acta Agriculturae Scandinavica, Section A — Animal Science* 60(3), 129–140. DOI: 10.1080/09064702.2010.523480.

Böhm, R. and Theelen, M.M.P. (2016) Outcome valence and externality valence framing in public good dilemmas. *Journal of Economic Psychology* 54, 151–163. DOI: 10.1016/j.joep.2016.04.003.

Boissy, A., Manteuffel, G., Jensen, M.B., Moe, R.O., Spruijt, B. *et al.* (2007) Assessment of positive emotions in animals to improve their welfare. *Physiology & Behavior* 92(3), 375–397. DOI: 10.1016/j.physbeh.2007.02.003.

Boulton, M.I., Wickens, A., Brown, D., Goode, J.A. and Gilbert, C.L. (1997) Prostaglandin F2α-induced nest-building in pseudopregnant pigs. I. Effects of environment on behaviour and cortisol secretion. *Physiology & Behavior* 62(5), 1071–1078. DOI: 10.1016/S0031-9384(97)00253-9.

Bracke, M.B.M. and Hopster, H. (2006) Assessing the importance of natural behavior for animal welfare. *Journal of Agricultural and Environmental Ethics* 19(1), 77–89. DOI: 10.1007/s10806-005-4493-7.

Braithwaite, V. and Huntingford, F. (2004) Fish and welfare: do fish have the capacity for pain perception and suffering? *Animal Welfare* 13, 87–92.

Brambell, F.W.R. (1965) *Report of the Technical Committee to Enquire into the Welfare of Animals kept under Intensive Livestock Husbandry Systems*. Her Majesty's Stationery Office, London.

Brooks, P.H. (2003) Group housing of sows – the European experience. *3rd London Swine Conference Proceedings: Maintaining Your Competitive Edge*, London, Ontario, Canada, pp. 37–60.

Buller, H. and Roe, E. (2012) Commodifying animal welfare. *Animal Welfare* 21(1), 131–135. DOI: 10.7120/096272812X13345905674042.

Burghardt, G.M., Ward, B. and Rosscoe, R. (1996) Problem of reptile play: Environmental enrichment and play behavior in a captive Nile soft-shelled turtle, Trionyx triunguis. *Zoo Biology* 15(3), 223–238. DOI: 10.1002/(SICI)1098-2361(1996)15:3<223::AID-ZOO3>3.0.CO;2-D.

Cain, P.J., Guy, J.G., Seddon, Y., Baxter, E.M. and Edwards, S.E. (2013) Estimating the economic impact of the adoption of novel non-crate sow farrowing systems in the UK. *International Journal of Agricultural Management* 2(2), 113–118. DOI: 10.5836/ijam/2013-02-06.

Carlsson, F., Frykblom, P. and Lagerkvist, C.J. (2007) Consumer willingness to pay for farm animal welfare: mobile abattoirs versus transportation to slaughter. *European Review of Agricultural Economics* 34(3), 321–344. DOI: 10.1093/erae/jbm025.

Carlstead, K., Mench, J.A., Meehan, C. and Brown, J.L. (2013) An epidemiological approach to welfare research in zoos: the elephant welfare project. *Journal of Applied Animal Welfare Science* 16(4), 319–337. DOI: 10.1080/10888705.2013.827915.

Carson, R. (1962) *Silent Spring*. Houghton Mifflin, Boston, Massachusetts.

Christensen, T., Denver, S. and Sandøe, P. (2019) How best to improve farm animal welfare? Four main approaches viewed from an economic perspective. *Animal Welfare* 28(1), 95–106. DOI: 10.7120/09627286.28.1.095.

Christensen, T., Lawrence, A., Lund, M., Stott, A. and Sandøe, P. (2012) How can economists help to improve animal welfare? *Animal Welfare* 21(1), 1–10. DOI: 10.7120/096272812X13345905673449.

Clark, B., Stewart, G.B., Panzone, L.A., Kyriazakis, I. and Frewer, L.J. (2016) A systematic review of public attitudes, perceptions and behaviours towards production diseases associated with farm animal welfare. *Journal of Agricultural and Environmental Ethics* 29(3), 455–478. DOI: 10.1007/s10806-016-9615-x.

Dawkins, M. (1980) *Animal Suffering: The Science of Animal Welfare*. Oxford University Press, Oxford, UK.

Dawkins, M.S. (1993) *Through Our Eyes Only. The Search for Animal Consciousness*. W.H. Freeman Spektrum, New York.

Dawkins, M.S. (2006) A user's guide to animal welfare science. *Trends in Ecology & Evolution* 21(2), 77–82. DOI: 10.1016/j.tree.2005.10.017.

Dawkins, M.S. (2016) Niko Tinbergen and W.H. Thorpe: very different answers to the question of animal awareness. In: Dwyer, C.M., Haskell, M.J. and Sandilands, V. (eds) *Proceedings of the 50th Congress of the International Society for Applied Ethology 12-15th July, 2016*. Edinburgh, UK, p. 70.

D'Eath, R.B. and Turner, S.P. (2009) The natural behaviour of the pig. In: Marchant-Forde, J.N. (ed.) *The Welfare of Pigs*. Springer Netherlands, Dordrecht, pp. 13–45.

D'Eath, R., Conington, J., Lawrence, A., Olsson, I. and Sand, P. (2010) Breeding for behavioural change in farm animals: practical, economic and ethical considerations. *Animal Welfare* 19, 17–27.

Defra (2004) *Animal Health and Welfare Strategy for Great Britain*. Defra Publications, London.

Defra (2006) Explanatory notes to animal welfare act 2006. Available at: www.legislation.gov. uk/ukpga/2006/45/notes/division/7/1/1 (accessed 1 April 2019).

Defra (2017) Animal welfare. Available at: www.gov.uk/guidance/animal-welfare (accessed 1 April 2019).

Defra (2019a) Cruelty to animals act 1876. Available at: www.legislation.gov.uk/ukpga/Vict/ 39-40/77/enacted (accessed 1 April 2019).

Defra (2019b) Defra, UK – science search. Available at: http://randd.defra.gov.uk/Default. aspx?Menu=Menu&Module=Detail&Completed=0&FOSID=3 (accessed 1 April 2019).

de Lauwere, C., van Asseldonk, M., van 't Riet, J., de Hoop, J. and ten Pierick, E. (2012) Understanding farmers' decisions with regard to animal welfare: the case of changing to group housing for pregnant sows. *Livestock Science* 143(2–3), 151–161. DOI: 10.1016/j. livsci.2011.09.007.

Douglas, C., Bateson, M., Walsh, C., Bédué, A. and Edwards, S.A. (2012) Environmental enrichment induces optimistic cognitive biases in pigs. *Applied Animal Behaviour Science* 139(1–2), 65–73. DOI: 10.1016/j.applanim.2012.02.018.

Duncan, I.J.H. (2006) The changing concept of animal sentience. *Applied Animal Behaviour Science* 100(1–2), 11–19. DOI: 10.1016/j.applanim.2006.04.011.

Duncan, I.J.H. and Wood-Gush, D.G.M. (1971) Frustration and aggression in the domestic fowl. *Animal Behaviour* 19(3), 500–504. DOI: 10.1016/S0003-3472(71)80104-5.

Edgar, J.L., Mullan, S.M., Pritchard, J.C., McFarlane, U.J.C. and Main, D.C.J. (2013) Towards a 'Good Life' for farm animals: development of a resource tier framework to achieve positive welfare for laying hens. *Animals* 3(3), 584–605. DOI: 10.3390/ani3030584.

Elwood, R.W. and Adams, L. (2015) Electric shock causes physiological stress responses in shore crabs, consistent with prediction of pain. *Biology Letters* 11(11), 20150800–20150803. DOI: 10.1098/rsbl.2015.0800.

European Commission (2012) *Communication from the Commission to the European Parliament, The Council and The European Economic and Social Committee on the European Union Strategy for the Protection and Welfare of Animals 2012–2015*. European Union, Brussels.

European Union (1997) Treaty of Amsterdam amending the Treaty on European Union, the treaties establishing the European Communities and certain related acts. *Office for Official Publications of the European Communities*, Bernan Associates [distributor], Luxembourg.

European Union (2007) Treaty of Lisbon: amending the Treaty on European Union and the Treaty establishing the European community. *Official Journal of the European Union*.

Available at: http://publications.europa.eu/resource/cellar/688a7a98-3110-4ffe-a6b3-8972d8445325.0007.01/DOC_19 (accessed 1 April 2019).

FAWC (1979) Farm animal welfare Council press statement. Available at: https://webarchive.nationalarchives.gov.uk/20121010012428/http://www.fawc.org.uk/pdf/fivefree-doms1979.pdf (accessed 1 April 2019).

FAWC (2005) *Report on the Welfare Implications of Farm Assurance Schemes*. Farm Animal Welfare Council, London, pp. 1–72.

FAWC (2008) *Opinion on Policy Instruments for Protecting and Improving Farm Animal Welfare*. Farm Animal Welfare Council, London, pp. 1–20.

FAWC (2009) *Farm Animal Welfare in Great Britain: Past, Present and Future*. Farm Animal Welfare Council, London, pp. 1–70.

Fox, J.A., Hayes, D.J. and Shogren, J.F. (2002) Consumer preferences for food irradiation: how favorable and unfavorable descriptions affect preferences for irradiated pork in experimental auctions. *Journal of Risk and Uncertainty* 24(1), 75–95. DOI: 10.1023/A:1013229427237.

Fraser, D., Weary, D.M., Pajor, E.A. and Milligan, B.N. (1997) A scientific conception of animal welfare that reflects ethical concerns. *Animal Welfare* 6, 187–205.

Frey, U.J. and Pirscher, F. (2018) Willingness to pay and moral stance: the case of farm animal welfare in Germany. *Plos One* 13(8), e0202193. DOI: 10.1371/journal.pone.0202193.

Gigerenzer, G. and Gaissmaier, W. (2011) Heuristic decision making. *Annual Review of Psychology* 62(1), 451–482. DOI: 10.1146/annurev-psych-120709-145346.

Gill, V. (2018) Crows reveal foundation of technology. BBC Science and Environment. Available at: https://www.bbc.com/news/science-environment-42655705 (accessed 1 April 2019).

Green, L.E., Hedges, V.J., Schukken, Y.H., Blowey, R.W. and Packington, A.J. (2002) The impact of clinical lameness on the milk yield of dairy cows. *Journal of Dairy Science* 85(9), 2250–2256. DOI: 10.3168/jds.S0022-0302(02)74304-X.

Griffin, D.R. (2013) *Animal Minds: Beyond Cognition to Consciousness*. University of Chicago Press, Chicago, Illinois.

Guy, J.H., Cain, P.J., Seddon, Y.M., Baxter, E.M. and Edwards, S.A. (2012) Economic evaluation of high welfare indoor farrowing systems for pigs. *Animal Welfare* 21(1), 19–24. DOI: 10.7120/096272812X13345905673520.

Hansen, K.E. and Curtis, S.E. (1980) Prepartal activity of sows in stall or pen. *Journal of Animal Science* 51(2), 456–460. DOI: 10.2527/jas1980.512456x.

Hansson, H. and Johan Lagerkvist, C. (2012) Measuring farmers' attitudes to animal welfare and health. *British Food Journal* 114(6), 840–852. DOI: 10.1108/00070701211234363.

Harding, E.J., Paul, E.S. and Mendl, M. (2004) Cognitive bias and affective state. *Nature* 427(6972), 312. DOI: 10.1038/427312a.

Harrison, R. (1964) *Animal Machines: The New Factory Farming Industry*. Vincent Stuart, London.

Hernandez-Mendo, O., von Keyserlingk, M.A.G., Veira, D.M. and Weary, D.M. (2007) Effects of pasture on lameness in dairy cows. *Journal of Dairy Science* 90(3), 1209–1214. DOI: 10.3168/jds.S0022-0302(07)71608-9.

House of Commons (2006) *Animal Welfare Act*. The Stationery Office Ltd., London.

Hughes, B.O. and Black, A.J. (1974) The effect of environmental factors on activity, selected behaviour patterns and 'fear' of fowls in cages and pens. *British Poultry Science* 15(4), 375–380. DOI: 10.1080/00071667408416121.

Hughes, B.O. and Duncan, I.J.H. (1988) The notion of ethological 'need', models of motivation and animal welfare. *Animal Behaviour* 36(6), 1696–1707. DOI: 10.1016/S0003-3472(88)80110-6.

Hume, D. (1987) *Essays - Moral, Political and Literary*, 2nd revised edn. Liberty Fund Inc, Indianapolis, Indiana.

Jarvis, S., Lawrence, A.B., McLean, K.A., Deans, L.A., Chirnside, J. *et al.* (1997) The effect of environment on behavioural activity, ACTH, (β-endorphin and cortisol in pre-farrowing gilts. *Animal Science* 65(3), 465–472. DOI: 10.1017/S1357729800008663.

Jarvis, S., McLean, K.A., Calvert, S.K., Deans, L.A., Chirnside, J. *et al.* (1999) The responsiveness of sows to their piglets in relation to the length of parturition and the involvement of endogenous opioids. *Applied Animal Behaviour Science* 63(3), 195–207. DOI: 10.1016/S0168-1591(99)00013-1.

Johansson-Stenman, O. (2018) Animal welfare and social decisions: is it time to take Bentham seriously? *Ecological Economics* 145, 90–103. DOI: 10.1016/j.ecolecon.2017.08.019.

Jones, R.C. (2013) Science, sentience, and animal welfare. *Biology & Philosophy* 28(1), 1–30. DOI: 10.1007/s10539-012-9351-1.

Kerr, S.G.C., Wood-Gush, D.G.M., Moser, H. and Whittemore, C.T. (1988) Enrichment of the production environment and the enhancement of welfare through the use of the Edinburgh Family Pen system of pig production. *Research and Development in Agriculture (UK)*.

Key, B. (2015) Fish do not feel pain and its implications for understanding phenomenal consciousness. *Biology & Philosophy* 30(2), 149–165. DOI: 10.1007/s10539-014-9469-4.

Künzl, C. and Sachser, N. (1999) The behavioral endocrinology of domestication: a comparison between the domestic guinea pig (*Cavia apereaf. porcellus*) and its wild ancestor, the Cavy (*Cavia aperea*). *Hormones and Behavior* 35(1), 28–37. DOI: 10.1006/hbeh.1998.1493.

Lagerkvist, C.J. and Hess, S. (2010) A meta-analysis of consumer willingness to pay for farm animal welfare. *European Review of Agricultural Economics* 38(1), 55–78. DOI: 10.1093/erae/jbq043.

Lassen, J., Sandøe, P. and Forkman, B. (2006) Happy pigs are dirty! – conflicting perspectives on animal welfare. *Livestock Science* 103(3), 221–230. DOI: 10.1016/j.livsci.2006.05.008.

Lawrence, A.B. (2008a) Applied animal behaviour science: past, present and future prospects. *Applied Animal Behaviour Science* 115(1–2), 1–24. DOI: 10.1016/j.applanim.2008.06.003.

Lawrence, A.B. (2016) Applied animal behaviour science and animal welfare: seeking the best balance between our science and its application. In: Brown, J., Seddon, Y. and Appleby, M. (eds) *Animals and Us; 50 Years and More of Applied Ethology*. Wageningen Academic Publishers, Wageningen, The Netherlands,, pp. 111–120.

Lawrence, A.B. (2008b) What is animal welfare. In: Branson, E. (ed.) *Fish Welfare*. Blackwell Publishing, Oxford, UK, pp. 7–18.

Lawrence, A.B. and Stott, A.W. (2009) Profiting from animal welfare: an animal-based perspective. *Journal of the Royal Agricultural Society of England* 170, 40–47.

Lawrence, A.B., Conington, J. and Simm, G. (2004) Breeding and animal welfare: practical and theoretical advantages of multi-trait selection. *Animal Welfare* 13, 191–196.

Lawrence, A.B., Newberry, R.C. and Špinka, M. (2018) Positive welfare: what does it add to the debate over pig welfare? In: Špinka, M. (ed.) *Advances in Pig Welfare, Herd and Flock Welfare*. Woodhead Publishing, Cambridge, UK, pp. 415–444.

Lawrence, A.B., Petherick, J.C., McLean, K.A., Deans, L.A., Chirnside, J. *et al.* (1994) The effect of environment on behaviour, plasma cortisol and prolactin in parturient sows. *Applied Animal Behaviour Science* 39(3-4), 313–330. DOI: 10.1016/0168-1591(94)90165-1.

Leach, K.A., Whay, H.R., Maggs, C.M., Barker, Z.E., Paul, E.S. *et al.* (2010) Working towards a reduction in cattle lameness: 1. understanding barriers to lameness control on dairy farms. *Research in Veterinary Science* 89(2), 311–317. DOI: 10.1016/j.rvsc.2010.02.014.

Lepage, O., Øverli, Ø., Petersson, E., Järvi, T. and Winberg, S. (2000) Differential stress coping in wild and domesticated sea trout. *Brain, Behavior and Evolution* 56(5), 259–268. DOI: 10.1159/000047209.

Macphail, E.M. (1998) *The Evolution of Consciousness*. Oxford University Press, New York.

Mather, J.A. and Anderson, R.C. (2007) Ethics and invertebrates: a cephalopod perspective. *Diseases of Aquatic Organisms* 75(2), 119–129. DOI: 10.3354/dao075119.

Mathews, A. and Mackintosh, B. (1998) A cognitive model of selective processing in anxiety. *Cognitive Therapy and Research* 22(6), 539–560. DOI: 10.1023/A:1018738019346.

Mathis, K. and Steffen, A.D. (2015) From rational choice to behavioural economics. In: Mathis, K. (ed.) *European Perspectives on Behavioural Law and Economics*. Springer International Publishing, Cham, pp. 31–48.

Mazeaud, M.M., Mazeaud, F. and Donaldson, E.M. (1977) Primary and secondary effects of stress in fish: some new data with a general review. *Transactions of the American Fisheries Society* 106(3), 201–212. DOI: 10.1577/1548-8659(1977)106<201:PASEOS>2.0.CO;2.

McCulloch, S.P. (2013) A critique of FAWC's Five Freedoms as a framework for the analysis of animal welfare. *Journal of Agricultural and Environmental Ethics* 26(5), 959–975. DOI: 10.1007/s10806-012-9434-7.

McInerney, J.P. (1994) Animal Welfare: An Economic Perspective. *Occasional Paper – Department of Agricultural Economics and Management*. University of Reading, UK.

McMullen, S. (2016) The ethical logic of economics. In: McMullen, S. (ed.) *Animals and the Economy, The Palgrave Macmillan Animal Ethics Series*. Palgrave Macmillan, London, pp. 27–44.

Mellor, D. (2016) Moving beyond the "Five Freedoms" by updating the "Five Provisions" and introducing aligned "Animal Welfare Aims". *Animals* 6(10), 59. DOI: 10.3390/ani6100059.

Mench, J.A., Sumner, D.A. and Rosen-Molina, J.T. (2011) Sustainability of egg production in the United States--the policy and market context. *Poultry Science* 90(1), 229–240. DOI: 10.3382/ps.2010-00844.

Mendl, M., Burman, O.H.P. and Paul, E.S. (2010) An integrative and functional framework for the study of animal emotion and mood. *Proceedings of the Royal Society B: Biological Sciences* 277(1696), 2895–2904. DOI: 10.1098/rspb.2010.0303.

Mendl, M., Paul, E.S. and Chittka, L. (2011) Animal behaviour: emotion in invertebrates? *Current Biology* 21(12), R463–R465. DOI: 10.1016/j.cub.2011.05.028.

Myers, D.G. and Diener, E. (1995) Who is happy? *Psychological Science* 6(1), 10–19. DOI: 10.1111/j.1467-9280.1995.tb00298.x.

Nelson, X.J. and Fijn, N. (2013) The use of visual media as a tool for investigating animal behaviour. *Animal Behaviour* 85(3), 525–536. DOI: 10.1016/j.anbehav.2012.12.009.

OIE (2019) Animal welfare at a glance. Available at: http://www.oie.int/en/animal-welfare/animal-welfare-at-a-glance/ (accessed 1 April 2019).

Radford, M. (2001) *Animal Welfare Law in Britain: Regulation and Responsibility*, 1st edn. Oxford University Press, Oxford, UK.

Ratner, R.K., Soman, D., Zauberman, G., Ariely, D., Carmon, Z. *et al.* (2008) How behavioral decision research can enhance consumer welfare: from freedom of choice to paternalistic intervention. *Marketing Letters* 19(3–4), 383–397. DOI: 10.1007/s11002-008-9044-3.

Rauw, W.M., Kanis, E., Noordhuizen-Stassen, E.N. and Grommers, F.J. (1998) Undesirable side effects of selection for high production efficiency in farm animals: a review. *Livestock Production Science* 56(1), 15–33. DOI: 10.1016/S0301-6226(98)00147-X.

Ribeiro, S.D.A. (2017) EU platform on animal welfare. Food Safety – European Commission. Available at: https://ec.europa.eu/food/animals/welfare/eu-platform-animal-welfare_en (accessed 1 April 2019).

Rose, J.D., Arlinghaus, R., Cooke, S.J., Diggles, B.K., Sawynok, W. *et al.* (2014) Can fish really feel pain? *Fish and Fisheries* 15(1), 97–133. DOI: 10.1111/faf.12010.

Rutherford, K.M.D., Donald, R.D., Lawrence, A.B. and Wemelsfelder, F. (2012) Qualitative behavioural assessment of emotionality in pigs. *Applied Animal Behaviour Science* 139(3–4), 218–224. DOI: 10.1016/j.applanim.2012.04.004.

Sainsbury, A.W., Bennett, P.M. and Kirkwood, J.K. (1995) The welfare of free-living wild animals in Europe: harm caused by human activities. *Animal Welfare* 4, 183–206.

Sandoe, P., Nielsen, B.L., Christensen, L.G. and Sorensen, P. (1999) Staying good while playing God – the ethics of breeding farm animals. *Animal Welfare* 8, 313–328.

Serpell, J. (2004) Factors influencing human attitudes to animals and their welfare. *Animal Welfare* 13, S145–S151.

Simm, G. (1998) *Genetic Improvement of Cattle and Sheep*. Farming Press, Ipswich, UK.

Skarstad, G.A., Terragni, L. and Torjusen, H. (2007) Animal welfare according to Norwegian consumers and producers: definitions and implications. *International Journal of Sociology of Food and Agriculture* 15, 74–90.

Sperry, R.W. (1993) The impact and promise of the cognitive revolution. *American Psychologist* 48(8), 878–885. DOI: 10.1037/0003-066X.48.8.878.

Stolba, A. and Wood-Gush, D.G.M. (1984) The identification of behavioural key features and their incorporation into a housing design for pigs. *Annales De Recherches Veterinaires. Annals of Veterinary Research* 15, 287–299.

Stolba, A. and Wood-Gush, D.G.M. (1989) The behaviour of pigs in a semi-natural environment. *Animal Science* 48(2), 419–425. DOI: 10.1017/S0003356100040411.

Te Velde, H., Aarts, N. and Van Woerkum, C. (2002) Dealing with ambivalence: farmers' and consumers' perceptions of animal welfare in livestock breeding. *Journal of Agricultural and Environmental Ethics* 15(2), 203–219. DOI: 10.1023/A:1015012403331.

Thaler, R.H. and Sunstein, C.R. (2008) *Nudge: Improving Decisions about Health, Wealth, and Happiness*. Yale University Press, New Haven, Connecticut.

Thornton, P.K. (2010) Livestock production: recent trends, future prospects. *Philosophical Transactions of the Royal Society B: Biological Sciences* 365(1554), 2853–2867. DOI: 10.1098/rstb.2010.0134.

Toma, L., Ashworth, C.J. and Stott, A.W. (2008) A partial equilibrium model of the linkages between animal welfare, trade and the environment in Scotland. In: *109th Seminar of European Association of Agricultural Economists, November 20–21, 2008*. Viterbo, Italy.

Tversky, A. and Kahneman, D. (1981) The framing of decisions and the psychology of choice. *Science* 211(4481), 453–458. DOI: 10.1126/science.7455683.

Van Horne, P.L.M. and Achterbosch, T.J. (2008) Animal welfare in poultry production systems: impact of EU standards on world trade. *World's Poultry Science Journal* 64(1), 40–52. DOI: 10.1017/S0043933907001705.

Vestergaard, K. (1982) Dust-bathing in the domestic fowl — diurnal rhythm and dust deprivation. *Applied Animal Ethology* 8(5), 487–495. DOI: 10.1016/0304-3762(82)90061-X.

Vigors, B. (2018) Reducing the consumer attitude–behaviour gap in animal welfare: The potential role of 'nudges'. *Animals* 8(12), 232. DOI: 10.3390/ani8120232.

Wall, E., Simm, G. and Moran, D. (2010) Developing breeding schemes to assist mitigation of greenhouse gas emissions. *Animal* 4(3), 366–376. DOI: 10.1017/S175173110999070X.

Wemelsfelder, F. (2012) A science of friendly pigs: Carving out a conceptual space for addressing animals as sentient beings. In: Birke, L. and Hockenhull, J. (eds) *Crossing Boundaries*. Brill, Leiden and Boston, pp. 223–249.

Wemelsfelder, F., Hunter, T.E.A., Mendl, M.T. and Lawrence, A.B. (2001) Assessing the 'whole animal': a free choice profiling approach. *Animal Behaviour* 62(2), 209–220. DOI: 10.1006/anbe.2001.1741.

Woods, A. (2004) The construction of an animal plague: foot and mouth disease in nineteenth-century Britain. *Social History of Medicine* 17(1), 23–39. DOI: 10.1093/shm/17.1.23.

Woods, A. (2012) From cruelty to welfare: the emergence of farm animal welfare in Britain, 1964–71. *Endeavour* 36(1), 14–22. DOI: 10.1016/j.endeavour.2011.10.003.

Yeates, J.W. and Main, D.C.J. (2008) Assessment of positive welfare: a review. *The Veterinary Journal* 175(3), 293–300. DOI: 10.1016/j.tvjl.2007.05.009.

2 Farm Animal Welfare: Do Free Markets Fail to Provide It?

CARMEN HUBBARD,[1]* BETH CLARK,[1] AND DAVID HARVEY[2]

[1]Centre for Rural Economy, School of Natural and Environmental Sciences, Newcastle University, Newcastle, United Kingdom
[2]Emeritus Professor of Agricultural Economics, Newcastle University, Newcastle, United Kingdom

'… it isn't economics if it isn't about choice.'

Krugman and Welles (2015), *Economics*

'… regardless of whether humans eat them or not, farm animals are dependent on humans and their choices.'

Norwood and Lusk (2011b), *Compassion by the Pound*

Summary

Animal welfare is often claimed to be a 'public good', i.e. requiring government intervention and legislation to ensure that animal welfare is respected. In other words, markets, on their own, cannot be relied on to deliver socially acceptable animal welfare. In fact, the issues surrounding animal welfare are more complex and subtle than this. This chapter first explains the general features of public goods, as defined and recognized in economics (Section 2). It then turns to the specific case of animal welfare (Section 3) and explains that outlawing cruelty to animals is clearly a genuine public good, but improving animal welfare can only be achieved by reflecting consumers' willingness to pay for better animal welfare production. However, there is a clear disconnect between citizens' apparent concerns about animal welfare and their exhibited willingness to pay for better animal welfare. Does this imply a clear market failure? This apparent failure is examined with the aid of a thought experiment, and identifies the nature of the problem – a combination of information and communication deficiencies with peoples' limited availability of time, resources and motivation to attend to all social issues with each and every purchase decision. The underlying problem is one of

*Corresponding author: carmen.hubbard@newcastle.ac.uk

consumption externality – other peoples' consumption decisions affect my/your assessment of our own welfare – since farmed animal welfare depends on peoples' consumption decisions.

2.1 Introduction

In general, animal welfare relates to the wellbeing of animals or their quality of life (Scott *et al.*, 2001). However, it is a very complex concept, both scientifically and morally (Duncan and Fraser, 1997; Fraser, 1999; Sandøe *et al.*, 2003; Fraser and Weary, 2004; Buller and Morris, 2003; Duncan, 2005; Hubbard *et al.*, 2007), and it 'can mean different things to different people' (Hewson, 2003). Its intricacy lies within the interdisciplinary nature of the animal welfare science and, more importantly, in 'its link to humans and their understanding of animals, especially in relation to feelings, needs and natural behaviour' (Hubbard and Scott, 2011, p. 79).

For the larger public, it tends to convey animals ranging free in green fields. This disregards the fact that it is entirely possible for poor welfare standards, neglect and disease to be present in free-range or outdoor production systems, while so-called 'intensive systems' may provide a high-quality environment and stockmanship. Research tells us (e.g. Clark *et al.*, 2016; Spooner *et al.*, 2014; Cornish *et al.*, 2016) that consumers often have little connection with modern agricultural production, resulting in a growth of misconceptions and an unfortunate readiness to attach human values to animal behaviour. At the same time, while a significant proportion say that protecting animal welfare is very important (e.g. 57% of respondents in the Eurobarometer survey of 2015) and that they would pay a premium for 'high(er)' animal welfare, shoppers do not always translate such intentions into purchasing decisions. Where they do, it may well be linked to perceived superiority of quality and taste, rather than purely their concerns about animal welfare. Some sectors of the population do seem to show a preference for labels specifying high welfare, particularly in relation to certain species, such as chickens, when these have been featured heavily in the media. The recent trend for vegetarianism, veganism and 'flexitarianism' (YouGov, 2019) is driven by human health and environmental concerns, and animal welfare, and is heavily popularized through extensive media and social media coverage, e.g. EATLancet 2019 report. Thus, all too often, the information upon which consumers rely continues to be a patchwork of potentially ill-informed sources. More labelling does not seem to be the answer (Uehleke and Hüttel, 2018). There is a growing disconnect between the public and the production systems their animal products come from (Clark *et al.*, 2016, 2019; Rothgerber and Rothberger, 2014). Even well-regulated accreditation schemes can be confusing and consumers do not always understand their labelling systems, or what standards they seek to ensure.

In response to the knowledge gap between consumers and production systems and the growing public concern regarding how livestock is reared, research into farm animal welfare has flourished. Animal welfare scientists have developed new approaches and criteria for evaluating and monitoring welfare at the farm level (e.g. Welfare Quality project). They have increasingly moved away from the conventional approach of evaluating the provision of resources necessary to ensure good welfare ('input assessments') and have instead focused on the use of animal-based measures of welfare ('outcome assessment') (European Food Safety Authority (EFSA), 2015, p. 5). However, the heterogeneity of farm species and the variety of production systems add to the challenges of defining and assessing what good farm animal welfare actually means. From an economic point of view, the lack of knowledge or relevant information (i.e. the asymmetric information problem, as labelled by economists) can lead to market failure; hence the need for government intervention (Norwood and Lusk, 2011b). In the case of farm animal welfare, this can be translated as the incapacity of competitive markets to achieve and deliver socially optimal levels of animal welfare (Harvey and Hubbard, 2013). But what does a socially optimal level mean, and does the market system fail in the case of farm animal welfare? If it does fail, who should be responsible for correcting it?

2.2 Defining Market Failure

Economics textbooks (e.g. Sloman, 2014; Sloman and Garrat, 2017; Krugman and Welles, 2015) teach us that a market system fails when the free market does not provide what society wants; or in economists' jargon, it fails to achieve 'social efficiency'. Hence, market failure occurs when markets fail to be efficient, i.e. to best allocate resources between alternative uses. This means that there are missed opportunities (in relation to production and consumption) that will make some people better off without making others worse off (Krugman and Welles, 2015). Social efficiency is achieved where no further net social gain can be made by producing more or less of a good or service. This (net social gain) will occur where marginal social cost (i.e. the additional cost incurred by society by producing/consuming one more unit of a good or service) equals marginal social benefit (the additional benefit gained by society by producing/consuming one more unit of a good/ service). Nevertheless, markets, *per se*, very rarely lead to social efficiency, so governments need to intervene.

There are several reasons why markets fail: market power (when a good or service is supplied by a single producer, the monopolist); so-called 'externalities' or third party/spill-over effects on others (e.g. costs or benefits of production or consumption experienced by society but not by the producers or consumers themselves); or because some goods (e.g. public goods), due to their nature, cannot be provided efficiently by markets (Sloman and Garrat, 2017; Krugman and Welles, 2015).

Specifically, markets work effectively because of two characteristics: property (ownership) rights and prices. The latter are the most important economic signals for efficient markets; that is 'because they [prices] convey the essential information about other people's cost and their willingness to pay' (Krugman and Welles, 2015, p. 123). Thus, markets depend on prices signalling accurately both willingness to pay from consumers and users, and willingness to continue providing from suppliers (firms); otherwise markets cannot balance demand (willingness to pay) with supply (willingness to provide). Moreover, markets depend on prices reflecting the social, rather than the private, costs and benefits. When the full social costs (costs to society) and the full social benefits of economic activities (benefits to society) are different from the sum of the private costs and benefits (as expressed in market prices), then externalities arise. These take place when the activity of one person, household or firm causes a loss (external cost) or a gain (external benefit) to another, and where these losses or gains are uncompensated or unrewarded. If the gain or loss effects are compensated or rewarded, then we have a normal market transaction. Hence, externalities are those effects which are outside (external to) the market mechanisms of payments and contracts so they do not account for in-market transactions. If these (externalities) exist (air pollution is the classic example), then the market system will fail to take proper account of society's costs and benefits.[1]

Goods and services, in general, are defined by two key characteristics: excludability and rivalry in consumption. A good is *excludable* when its consumption requires payment (e.g. if you would like to consume a bar of chocolate you have to pay for it). If you do not pay then you are excluded from consuming it. A good is *rival in consumption* when the same unit of a good (e.g. bar of chocolate) cannot be consumed by more than one individual at the same time. Most of the goods and services that we buy on a day-to-day basis (e.g. bread, chocolate, milk, clothes, wine, washing machines and cars) are both excludable and rival in consumption. Economists label these as 'private goods', and free markets, in general, work efficiently in providing them. However, not all goods are excludable and rival in consumption. Some can be *non-excludable,* i.e. the consumption of the good cannot be prevented for those who do not pay for it (e.g. free water in shops/airports). Others are *non-rival in consumption,* i.e. more individuals can consume the same unit of good at the same time (e.g. TV shows) (Krugman and Welles, 2015). Hence, goods can be either excludable or non-excludable, rival or non-rival in consumption or a combination of these characteristics.

Goods which are *both* non-excludable and non-rival in consumption are defined as 'public goods'; the classic examples are lighthouses, street lights, street cleaning, pavements, flood control dams, sewage systems, disease prevention and national defence. These are goods that are socially desirable but privately unprofitable, i.e. large external benefits relative to private benefits, hence they cannot easily be provided by the market system. This is because the suppliers cannot guarantee to be paid for their efforts by the users or beneficiaries. The consequence is that the market can no longer be relied upon to generate a socially optimal allocation of resources to different forms of output, and hence cannot provide an optimal mix of end-products. There is also another class of goods (and especially services) provided from the public purse (though not principally because they are non-rival and non-excludable but because they might be under-consumed if not provided by government), which clearly overlaps with public goods. These are known as *merit goods*[2] (or services). Education and healthcare are classic examples and they are considered by the public through their governments to be sufficiently important (or necessary) to everyone that they should be provided to all, independently of ability to pay.[3]

One of the key issues encountered by goods which are non-excludable is the 'free-rider problem', that is when it is not possible to exclude other people from consuming a good or a service that someone has already bought (Sloman and Garrat, 2017). Thus, if the good (or service) is a 'public good', and everyone gets to enjoy it if it is provided, what incentive is there for any one individual to consent to paying their share of the costs of provision? Public goods thus suffer from this free-rider problem; with not enough reward being provided for their supply, so not enough of the public goods are provided. This leads to the under-provision of these goods and services and hence to market failure. As already noted above, the usual answer is for governments to provide these public goods, for example through tax revenues, to which everyone is required to contribute according to their means. However, there are other ways of providing public goods. As people also behave altruistically (for the good of society as well as, or even in competition with, their own self-interests) they can contribute voluntarily towards the provision of some public goods (e.g. private donations to support lighthouses in the UK or the Royal National Lifeboat Institution and volunteering for fire departments). So an alternative to the direct provision of public goods would be for the government to subsidize these voluntary efforts, thus encouraging more voluntary contributions.[4] Some individuals or firms can also supply public goods 'because those who produce them are able to make money in an indirect way … [e.g.] broadcast television, which in the USA is supported entirely by advertising' (Krugman and Welles, 2015, p. 493). However, as in the case of (positive) externalities, 'the marginal social benefit[5] of one or more unit of a public good is always greater than the individual marginal benefit to any one individual' (Krugman and Welles, 2015, p. 497), so there are insufficient incentives to provide a socially optimal level of that good. The solution requires that all the people interested in and benefiting from the public good are fully committed to the idea of collective action, and recognize that the solution will not work if people free-ride. But, leaving the provision of public goods to altruistic people or self-interested individuals or firms to provide an efficient amount of these goods, such as street lights or national defence, may prove to be difficult, hence the role of governmental intervention.

Although in theory the provision of public goods by governments may look straightforward, in practice it is more complicated as governments are confronted with the question of how much to provide of that public good. A cost-benefit analysis, which will allow for the estimation and comparison of social costs and social benefits of providing that good, might help governments to answer this question. But whereas estimating the social cost of supplying that good is easy/easier, estimating the (marginal) social benefit of a public good is very difficult. Some will suggest that willingness to pay

(WTP), i.e. the maximum amount an individual will pay to obtain (one more unit of) that good/service, can be employed to estimate (marginal) social benefit. But in the case of public goods, as people do not actually pay for them, the use of WTP to advise governments on the provision of public goods is controversial, and it calls for caution. That is because in the absence of incentives (and due to self-interest) individuals tend 'to overstate their true feelings when asked how much they desire a public good', so inflating the value of the goods (Krugman and Welles, 2015, p. 498). Additionally, by definition, externalities and public goods are not freely traded between people, so no one has any experience of trading these effects or benefits off against other things. In general, people put value on things or effects for three major reasons (McInerney, 2004a):

- *Use values:* the thing or effect is valuable because it is useful in providing either productive capacity or consumption benefits;
- *Option values*: things are valuable because they may become useful sometime in the future (or in different circumstances), so it is worth having them about as an insurance policy;
- *Existence value:* things are valuable in their own right, so it is nice to have them about.

None of these values, however, can be infinite or invaluable. Obtaining or preserving them always requires time, effort, inputs and resources, all of which could be used for other things; we have to make choices about what we do and what we produce or preserve, we cannot do everything. Choice automatically implies an *opportunity cost*, i.e. what we might have done instead. It will only be worthwhile to provide these things if the opportunity cost of providing them is lower than their values. Hence, if we do not think they are worth the effort, we will eventually stop providing for their preservation. However, unless we actually require people to put their money (or their own time and effort) where their mouth is, we will frequently get inconsistent and incoherent answers to the apparently simple question about how much a public good is worth. If we think someone else will pay, it makes sense for us to pretend or imagine that the thing is very valuable indeed. Once we are required to pay up to the value we think we place on the thing, we become very much more realistic about its worth in relative terms to the other things commanding our attention, resources and income. Economics (the market) cannot resolve these problems, except in very special circumstances. Solution requires collective action, i.e. governance – the long arm of the law is necessarily attached to the invisible hand of the market – not just to correct for market failure, but also to police the market system (e.g. outlaw theft, establish and police property rights and the laws of contract, and make judgements about relative equity of market outcomes).[6]

But how do all of the above relate to farm animal welfare? How do markets react to animal welfare? Is, indeed, animal welfare a public good, non-excludable and non-rival in consumption? If yes, is the free-riding problem big enough that in order to correct the market failure government intervention is required? How much are people actually willing to pay for it, so the government knows how much of it to provide and how much to spend making it available? Why don't people pay for improved animal welfare through the animal products they buy? What are the reasons for the apparent gap between votes (opinion poll surveys) in favour of improved animal welfare and the actual spending on improved AW products?

2.3   Do Markets Fail in the Case of Farm Animal Welfare (FAW)?[7]

As highlighted above, what constitutes *good* farm animal welfare and how it is measured is subject to debate, with no universally accepted definition. Furthermore, from an economic point of view, farm animal welfare is not a usual good (or service), so discussion regarding the potential for markets failing to provide it and the implications for government intervention through imposed legislation and regulations remains highly controversial. While animal welfare is often cited as a classic public good, Harvey and Hubbard (2013) and Mann (2005) consider that the public good argument needs substantial clarification, particularly in relation to how free markets affect animal welfare. In his seminal work, McInerney (2004a) stresses the importance of socio-economic choices/considerations in determining how farm animals are treated. In economic terms, animals are nothing else but a resource employed in food production. Hence, their value and importance are linked to what they contribute to economic output, or, more precisely, to their productivity or their 'use' (commercial) value (Fig. 2.1). However, society, with its social norms, moral responsibilities and individual preferences, adds additional value to animals. That is a 'non-use' or an ethical value which cannot be ignored.

However, as value is a human construct (or a reflection of our perceptions), 'animal welfare is … a subset of human welfare, the animals' preferences and wellbeing having relevance only to the extent that they are important to us' (McInerney, 2004a, p. 11). So 'the (use) value derived from

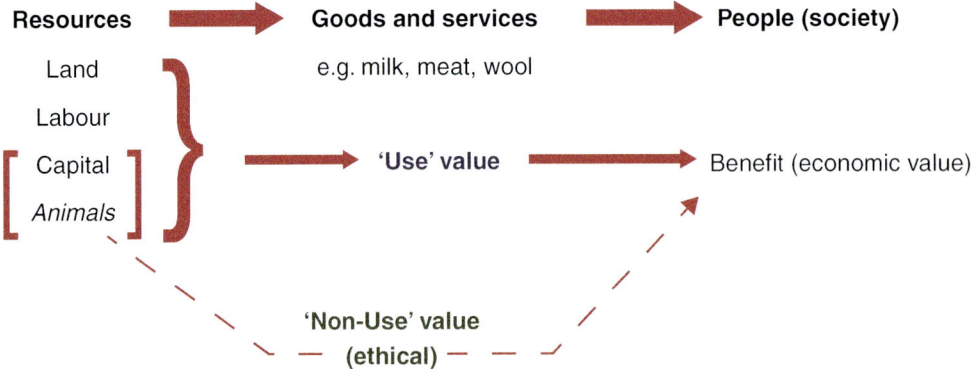

Fig. 2.1. The economic framework of livestock farming. (From McInerney, 2004a)

animals by virtue of their productivity is conditioned by the positive or negative (non-use) value attached to [human] perceptions of how well the animals themselves fared in the process' (McInerney, 2004b). It is this 'ethical', 'social' or 'non-use' value attached by humans and their perceptions of what means good (or bad) animal welfare that determine animal welfare, and it adds to the debate around the 'public good' argument. However, McInerney (2004a) stresses that, in practice, given its complexity and multi-dimensions, animal welfare is not a simple, clear binary choice about 'good' or 'bad' welfare. Hence, from an economic standpoint, the analysis should focus 'on assessing the value of *increments* [higher levels] of animal welfare, not the valuation attached to the overall welfare state of the animal(s). ... [That is] the recognition of value as a marginal concept – the worth or cost of acquiring something additional, not the total valuation of some particular quantity' (McInerney, 2004a).

In modern societies, social concerns for animal welfare are reflected in laws and regulations set up to provide (at least) a minimum level of standards for (farm) animal welfare. The European Union (EU) is at the forefront of promoting them, with the European Convention for the Protection of Animals Kept for Farming Purposes (1976) as the first EU Act related to farm animal welfare. The Council Directive 98/58/EC lays down the minimum standards for the protection of all farmed animals, and the Treaty of Amsterdam (European Union, 1997) recognized animals as sentient beings. In the USA, the Congress passes, annually, tens of pieces of legislation regarding animal welfare (Norwood and Lusk, 2011b). In some EU countries such as Sweden, Germany, The Netherlands and the UK, animal welfare standards are much higher relative to other EU Member States, with national legislation well beyond the minimum EU standards (particularly for some livestock sectors) (Schmid and Kilchsperger, 2011). But whereas in Sweden improved animal welfare standards and compliance with them is driven almost entirely by (governmental) legislation, in The Netherlands and the UK these are strongly promoted through market forces (Keeling *et al.*, 2012). However, delegation of responsibility for animal welfare by citizens to their governments (through tighter regulation) does not and cannot absolve citizens of responsibility for the consequences – *delegation does not mean abdication of responsibility.*

Decent animal welfare is often said to be a *public good*: once provided for one person, it is also provided for everyone (non-rival in consumption) and no one can be prevented from enjoying the benefits of improved animal welfare (non-excludable). Clearly, cruelty to animals is a *public bad*, at least as far as most modern societies are concerned. Many people are discomforted by the knowledge that animals are being cruelly treated in their society and take steps to ensure that this does not happen, usually through persuading their governments to outlaw the practices which discomfort them, and to take the necessary enforcement measures to ensure compliance with the prevention of cruelty law. Absence of cruelty (enforced regulation) is a clear 'public good' – once provided for one, it is provided for all. However, as discussed in Section 2, the regulation can only be judged to be 'good' if society, *as a whole,* considers the gains resulting from the regulation to more than offset any as-sociated costs and disadvantages. Taking the decision to introduce the regu-lation implies an underlying social cost/benefit calculation in favour of the regulation; otherwise how does government or society judge the net benefit on society's behalf? Since there is no (defensible) cost to outlawing cruelty – once a common definition of cruelty can be agreed – it is clearly in the social interest to outlaw cruelty. The difficulties arise over whether particular cur-rent practices (e.g. fox hunting, bull fighting, battery cages, gestation crates) are commonly agreed to be cruel. Clearly, those who pursue and support the practices disagree that they are unnecessarily cruel and will experience a loss as and when they are outlawed. The 'majority', however, will experience the public good benefit of knowing that the practices are no longer pursued. However, as Mann (2005) notes, at least some people experience consider-able discomfort knowing that animal welfare standards are not as high as they should be, over and above outlawed unnecessary cruelty, regardless of their own consumption (or not) of animal products.

The essence of the 'public good' argument in the case of animal welfare is that some people's own value of animal welfare is affected by what other people do and what they consume. In economic terms, this discomfort is a 'consumption externality' – an effect on an individual wellbeing generated by another's individual consumption. Since overall farm animal welfare de-pends on both the welfare of each animal and the number of animals, and as farm animals are kept to meet consumer and market demands (otherwise animal products will cease to exist), actual animal welfare (as an aggregate or average level of animal welfare over the whole farmed animal popula-tion) depends on the demand for animal products. These animal products (hence the welfare they imply) are clearly excludable and rival (*private*) goods. Nevertheless, the consumption externality is a form of market failure. Does willingness to pay adequately reflect this externality, or is government intervention required?

Much like the debate of what is good farm animal welfare, assessing public perceptions can be also highly subjective (Clark *et al.*, 2016). Some will argue that economics can provide more of an objective assessment of the farm animal welfare by putting forward individuals' views as stated preferences (Norwood and Lusk, 2011b) by employing choice modelling or willingness to pay (WTP) techniques. These are usually tied to a specific product and product attributes. It also enables trade-offs between different viewpoints, such as individuals and society, to be achieved (Norwood and Lusk, 2011b), and so it gives an indication as to how people think animals should be raised.

Economics could also provide an indication as to what *trade-offs* people are interested in. In this sense, in order to improve farm animal welfare, an individual must give up something else in return, but the question is how much (Norwood and Lusk, 2011a)? The answer to this question could be refraining from eating animal products altogether, eating less of an animal product or paying more money for them. A carefully designed willingness-to-pay study seeks to understand the choices that individuals would make with their own money; however, 'the economic approach also asserts that *individuals tend to be most happy when they are free to make their own choices in the marketplace'* (Norwood and Lusk, 2011b, p. 4).

Several reviews (e.g. Clark *et al.*, 2017; Lagerkvist and Hess, 2011; Cicia and Colantuoni, 2010; Lusk *et al.*, 2003) have explored and summarized consumers' WTP for food attributes, including high(er) animal welfare products. In spite of an apparently strong (marginal) willingness to pay for an animal welfare attribute (particularly for some species, e.g. chicken) that came out of these studies, research (e.g. Harper and Henson, 2001; McVittie *et al.*, 2006; Toma *et al.*, 2011) also shows that consumers' stated preferences/attitudes towards farm animal welfare do not always translate into purchase decisions. Hence, there is a considerable gap between the reported levels of willingness to pay and the actual sales of high(er) welfare products (Clark *et al.*, 2017; Baltzer, 2004). Therefore, the market for improved animal welfare products is regarded as niche rather than mainstream. In other words, it can be argued that the 'economic value attached to increments in [standards of] welfare can be treated as a private good which government has no responsibility to provide' (McInerney, 2004a). The purchase of high(er) welfare products creates private value and provides some form of benefit to those who care about farm animal welfare (McVittie *et al.*, 2006), be that from an ethical perspective, or from an indication to consumers of improved quality or sensory properties such as taste (Uehleke and Hüttel, 2018), safety or nutrition (Clark *et al.*, 2016). These benefits emphasize that farm animal welfare is primarily seen as a credence attribute (requiring consumers to believe the claim, which cannot be verified merely by experiencing consumption of the good), and consumers' actual purchasing behaviour contrasts significantly with the growing public concern regarding how farm animals are reared and treated, i.e. standards well beyond the minimum legislation. People, as citizens, profess substantial concern about animal welfare but do not 'put their money where their mouths are' when it comes to doing something about it. If more people were prepared to spend more money on improved animal welfare products, the market would be encouraged to respond by providing them, with the premium necessary to cover any additional costs.

Several reasons have been put forward for this disconnect between attitude (or 'voting intention') as citizens and actual behaviour (revealed preferences) as consumers. Clearly, consumers have a number of competing interests at the point-of-purchase. Subsequently, willingness to pay is likely to change depending on these interests, the information available and the context. As Kjærnes (2012) argues, the choices people make are often based on conflated assumptions about what is 'good enough' for themselves, the environment and farm animals; for example, an inference that organic food scores more highly on all counts. Cultural context is also very important; thus by conforming to social norms, people remove the need to be constantly reconsidering the morality of their choices (Kjaernes and Lavik, 2007 in Hubbard *et al.*, 2018). Moreover, when consumers are asked to spontaneously list the product attributes that they look for when making a food purchase, animal welfare is mentioned much less frequently or given lower priority. In addition, barriers to purchase such as high prices, lack of availability and poor labelling have all been listed as reasons why levels of concern do not match up to reported levels of WTP. A lack of knowledge about modern farming practices makes it difficult for consumers to express their preferences, especially when labelling is inconsistent, unclear or simply not present. As Hubbard *et al.* (2018) highlights:

> There is … a knowledge gap around farming production systems and very little (if any) research that links the accuracy of knowledge of farm animal welfare and the provision of information, to attitudes or consumer behaviour *per se* … [and] no direct research available that unpacks the relationship between knowledge of farmed species and consumer behaviour.

In addition to these conscious barriers, there are also subconscious reasons to purchasing improved animal welfare products based on moral disengagement (believing certain ethical standards do not apply to oneself), cognitive dissonance strategies (a state of psychological discomfort from inconsistencies between thoughts) or attitudinal ambivalence (having both positive and negative attitudes towards an object/behaviour without the accompanying mental discomfort). All three factors are thought to influence animal product consumption to some degree, through the diffusion or displacement of responsibility for animal welfare (Clark *et al.*, 2016) or the 'meat paradox' (Loughnan *et al.*, 2014), whereby individuals hold a simultaneous desire to treat animals well yet still eat them. In relation to meat, Graça *et al.* (2016) have identified a number of disengagement processes, namely: (i) means-ends justifications, i.e. pro-meat rationalizations; (ii) desensitization, i.e. becoming emotionally and cognitively desensitized to the death and suffering of animals used for food purposes; (iii) the denial of negative consequences of production and consumption; (iv) reduced perceived choice, i.e. framing alternatives as inaccessible and impractical; and (v) as aforementioned, diffused responsibility. Nevertheless, it is apparent that people as citizens do say they support animal welfare, and also assert that more could and should be done to improve animal welfare conditions, reflecting society's moral and ethical values (e.g. Bennett and Blaney, 2002).

Vanhonacker *et al.* (2007, p. 86) label this as the 'consumer–citizen duality'. The public pressure for legislation, improved standards and related government action on animal welfare is typically supposed to come from citizens rather than consumers (albeit they are the same people). Within the EU, there are few mechanisms through which citizens can vote directly for government action on animal welfare. For example, constituent pressure for government action on animal welfare comes from public attitude surveys (e.g. European Commission, 2007 and European Commission, 2016) and from indirect pressure on political representatives and assemblies from individual concerned citizens (including vegans and vegetarians) and advocacy groups (NGOs) seeking better animal standards (e.g. Compassion in World Farming). Clearly, some citizens are much more concerned than others, with those seriously enough concerned to devote time and resources to campaigning typically in the minority. There are, however, criticisms that such surveys do not account for framing bias (the tendency for people to agree with statements reflecting perceived social norms, e.g. Vanhonacker *et al.*, 2007); scaling bias (associated with the length of scale offered to respondents); and lack of any relativity or assessment of salience when asking about importance (relative to other perhaps more important personal, social and public issues). Additionally, there is only a limited literature on why these differences exist and their implications for market failure (e.g. Vanhonacker *et al.*, 2007; Harper and Henson, 2001; and also Hamilton *et al.*, 2003, with reference to pesticides).

2.4   WTP versus Actual Purchases for Animal Welfare: A Thought Experiment[8]

Suppose that we could obtain reliable panel data on: peoples' stated concerns about animal welfare (their votes in favour of improved animal welfare); their stated WTP for improved animal welfare products; and their actual spending on these products. By reliable data, we mean that these data would be very closely replicated in repeat surveys and observations, implying that they are collected from appropriate samples of the population, appropriately controlled and corrected for known biases and appropriately analysed to correct for confounding factors. Would we expect these three sets of data to say the same thing? More importantly, what would we expect these same people to say, as and when we confront them with the apparent disparities between their voting and WTP responses and their actual purchasing behaviour? The difficulties of conducting such an experiment in practice need not detain us here (though Andersen (2011) reports part of such a study, examining WTP and purchase behaviour). Both common sense and economics suggest that we should try to identify what, if anything, we would learn from such an experiment before committing valuable resources and effort to overcoming the very substantial difficulties in implementing the experiment in practice.

The potential reasons for discrepancies between a person's vote and their expressed WTP (elicited in a well designed choice experiment), and between their expressed WTP and their actual purchase behaviour, are closely analogous. Both the vote and the WTP are 'hypothetical' – expressions of personal (individual) judgement and intention – while the purchase behaviour is the exhibition of each person's (household's) actual spending on improved animal welfare.[9] Carson and Groves (2007, p. 206) recommend that researchers avoid 'the use of the word *hypothetical* in discussing preference questions, in favour of *consequential* and *inconsequential* to emphasize the conditions requisite for the application of economic theory'. Their strong point is that it is only possible to make logical conjectures about (and inferences from) peoples' responses to *consequential* questions. As these same authors (Carson and Groves, 2011) point out:

> For a question to be consequential, survey respondents need to believe, at least probabilistically, that their responses to the survey may influence some decision they care about. For consequential survey questions, economic

theory is relevant in terms of the incentives respondents face in answering the question. … For inconsequential decisions, any response is as good as any other response in terms of its influence on the respondent's utility level.

So, in order to persuade respondents to provide some economically meaningful response, our WTP survey needs to be seen as consequential in order for their responses to be 'incentive compatible'. In other words, our respondents have every incentive to respond strategically, if this appears sensible to them. They (we) can be expected to at least 'disguise' or be 'economical with' the truth – whatever that is – in their (our) responses, if doing so promotes their (our) chances of being better off as a result.

We also consider the potential reasons that might be offered by our conceptual respondents for the differences between their votes and their WTP, or their WTP and their actual spending. To structure the explanation of our thought experiment, we present the summary of the results, as an anatomy of market failure for animal welfare, in Fig. 2.2. Given that our illustrative respondents have already indicated that they are in favour of improved animal welfare ('voted' for it), how might we expect them to respond to a subsequent question about how much they are willing to pay (or how much improved animal welfare they actually purchase)?

The first possible answer (Fig. 2.2, 1) is 'nothing at all'. In other words, the response to the vote question, assuming it is treated as consequential, is intended to persuade others, or the authorities, to improve animal welfare, and to show support for such socially beneficial progress. This vote does not signal, however, that we are willing to or think we need to pay anything for this. A reason might well be that 'it is the government's job to safeguard animal welfare, not mine'. Alternatively, it might be argued that the market counts euros (or dollars) rather than votes, so cannot measure ethical

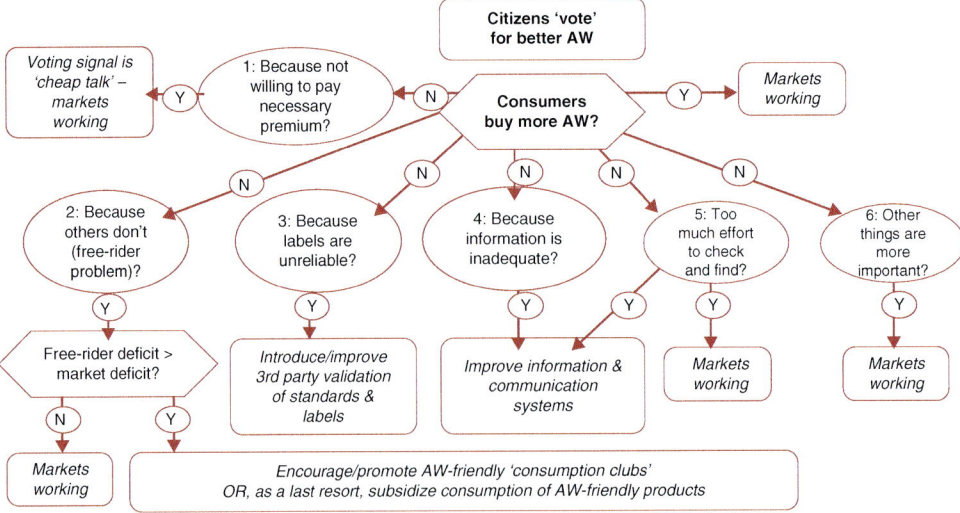

Fig. 2.2. Citizens versus consumers: anatomy of market failure for improved animal welfare.

issues properly, which is the government's job. Or our respondents might well object strongly to anybody consuming animal products and thus contributing to animal ill-fare, while certainly not being willing to pay anything for any animal product. Lusk (2011), for example, considers these arguments and suggests that developing a separate market in 'animal wellbeing units', where producers could respond directly to citizens' willingness to pay for improved animal welfare, might be a route through which, for instance, vegans and vegetarians could exert their influence. This is an ingenious economic suggestion to internalize an otherwise missing market, but would have very considerable implementation and operational costs. It would also, doubtless, meet indignant objections from deontological ethicists (as Lusk notes) who believe that it is completely unethical to trade animal wellbeing in any market. At least in a democracy the only course open to ethicists is to convert as many others as possible to their moral values, thereby reducing consumption of animal products and the associated abuse of their innate rights. Otherwise, these particular values can only be imposed throughout society through legislation, based on majority vote.

However, this perfectly reasonable response leads back to the conclusion that regulation is the best, if not the only, way of maintaining and improving animal welfare. This in turn has to solve the serious problem of how to make social decisions on the appropriate levels of animal welfare in the face of widely differing and diverse views and judgements about the appropriate levels (as noted above). Whatever our underlying reasons for this lack of WTP, which are doubtless extremely complex, it is 'cheap talk' to vote for something for which one is not prepared to pay (Andersen, 2011). Carson and Groves (2011) note, importantly, that the term 'cheap talk' originated in the game theory literature, which 'shows that talk is not "cheap" when it can influence the actions of others. What was thought of initially as a costless way of signalling … turns out to be quite consequential in the right circumstances.' If citizens do not exhibit or express any WTP for animal welfare-friendly products, markets cannot be expected to work, although cheap talk may be influential in the ways they work, if listened to and accepted by others. In any event, the fact that markets do not respond to attitudes or votes without the necessary inducement of payment, or that the market's many participants do not subscribe to your (or my) particular moral codes, cannot be described usefully as 'market failure'.

An alternative answer to the WTP question is, a lot, without necessarily being truthful, on the grounds that greater willingness-to-pay signals might bring forth an increased supply of animal welfare-friendly products. We might expect WTP approximations (at least for those expressed above zero) to generally overestimate actual WTP as revealed by purchase behaviour (Andersen, 2011), so long as the survey or choice experiment questions are treated as consequential. As Carson and Groves (2007, p. 188), put it:

> As long as there is any positive probability of wanting the new good at the stated price, the respondent should say 'Yes — would purchase.' The respondent's logic is that such an answer will encourage the chain to produce

the good, with the respondent able to decide later whether or not to purchase. Since increasing the respondent's choices in a desirable way increases utility (and hence social welfare), the optimal response is 'yes.'[10]

In this case, our thought experiment (as illustrated in Fig. 2.2) reverts to comparison between the large positive WTP and associated actual purchases of animal welfare-friendly products.

There are five major classes of substantive reasons that can be offered to explain the difference between votes and WTP, and subsequently between WTP and actual purchase behaviour. The first of these (Fig. 2.2, 2) is that (some) people would be willing to pay the premium if only they could be sure that their contribution to improved animal welfare would actually make a difference to animal welfare, and/or if they could be sure that other people, too, were willing to support improved animal welfare. My consumption of products from poor animal welfare production systems helps to perpetuate these poor systems, and adversely affects others who value animal welfare more highly than I do.[11] If I can be persuaded not to consume these products, others will feel better off, and I might well feel better simply as a consequence of their improved wellbeing. Thus, I might be more prepared to switch my consumption towards more animal welfare-friendly products (even at higher cost) if I could be sure that other people would also do so. But if not, there is a strong temptation to consider that my own efforts in favour of better animal welfare are too small to make any substantial difference, and hence not worth my effort and spending. These consequences are aspects of the free-rider problem, the root cause of potential market failure in the case of animal welfare. In total (summed over the relevant population), the difference between the sum of individual WTP and the sum of WTP conditional on other people also paying for animal welfare can be termed the *free-rider deficit*. In principle, if this deficit is greater than the amount (cost) necessary to encourage the supply chain to deliver the associated (additional) improvement in animal welfare (the *market deficit*), then society would be better off if the free-rider problem could be solved. If the free-rider deficit is not greater than the market deficit, then although there is a potential free-rider problem, it does not result in any market failure; society would not be better off by 'solving' the problem – the costs of doing so would outweigh the benefits, at least as measured by WTP backed up with actual spending.[12]

In practice, it is observed (e.g. Verbeke, 2009, p. 327) that some people do appreciate the social pressures and norms which encourage them to spend their own money on improving animal welfare – hence both implicitly recognizing and dealing with the free-rider problem. In effect, these people consider themselves part of a club or society in which the social pressures and norms are sufficient to persuade them to conform in spite of, rather than because of, their own self-interests. As societies spend more time and effort improving animal welfare, so people become more aware of the issues and more likely to respond to these growing social pressures. This 'involvement' of consumers with the products and their provenance, either directly or indirectly through social norms, can be improved, for some segments of the market, through better communication and information. This effectively encourages people to join the 'virtual' clubs of those concerned about animal welfare. In short, simple economics, which assumes everyone is purely self-interested and rational, suggests that the free-rider problem could be substantial. In the real world, in which very few are simple *homus economicus*, and many are responsive to their peers and social norms, the free-rider problem may not be as significant. Social norms and values can be cultivated and encouraged through active debate and provision of objective, disinterested information and validation services, which need to be provided through collective and collaborative action (governance), if not at public expense; hence 'everyone is responsible'. Measuring the free-rider deficit is, however, fraught with serious difficulty. It involves measuring willingness to pay as a precondition, and subsequently identifying the differences between this estimate with and without the condition that others are also both willing to and actually do pay their 'share'. Recently, Uehleke and Hüttel (2018) measured the free-rider deficit. Using choice experiments, they examined the consumers' stated demand for a high(er) farm animal welfare label in Germany, in an individual and a collective decision situation, estimating the size of the potential externality in a given market (e.g. labelled chicken and pork). They found significantly higher support rates for animal welfare-labelled products under the collective scenario, i.e. a hypothetical referendum for a mandatory label, hence indicating the presence of a significant externality in the market for labelled meat. However, as the authors acknowledge, their experiment focused on the relative market share or support rates rather than on the absolute WTP, and their results are 'not linked to economic consequences ... therefore might suffer from hypothetical bias' (Uehleke and Hüttel, 2018). This is not surprising, as since the questions eliciting peoples' WTP are hypothetical, the answers are always subject to a sort of hypothetical bias. People do not actually do what they say they will do, because 'other things are not (and never are) equal'. Some analysts have gone as far as to argue that this hypothetical bias is sufficiently strong as to render all such 'contingent valuation' exercises effectively meaningless (e.g. Diamond and Hausman, 1994). Others, e.g. Blamey *et al.*, 1995, argue that people express different preferences for public (or club) goods (such as improved animal welfare) than for private (normal) goods. Further, these two sorts of preferences (and the associated decisions) are conceptually distinct and non-commensurate – termed the 'citizen hypothesis'. However, Curtis and McConnell (2002), using data involving preferences for deer-culling programmes in the USA, demonstrate that the citizen hypothesis is observationally equivalent to the 'standard' self-interested hypothesis augmented to include altruism. As Curtis and McConnell (2002) note: 'The citizen hypothesis is not empirically testable. It is a maintained hypothesis because the citizen hypothesis concerns the individuals' underlying motives and these motives are never conclusively revealed in actual behaviour or survey responses' (p. 72). They conclude (p. 82): 'At least in the case we have studied, there is no difference in the WTP for respondents who could be reasonably classified as citizens and consumers.' Uehleke and Hüttel (2018) also stress that they 'cannot observe the motives why subjects do not choose the labelled meat', hence their focus on the differences in individual and collective WTP.

There are, however, four other substantive reasons for discrepancy between votes, WTP and actual spending, other than 'cheap talk' and 'free rider'. A third reason (Fig. 2.2, 3) is that the labelling of animal welfare-friendly products is not sufficiently obvious or reliable to attract consumers or convince them that their additional spending will encourage improved welfare. A fourth reason (Fig. 2.2, 4) is closely related, i.e. the information available to consumers about animal welfare and the improved standards used in producing some products is insufficient (or too contradictory or confusing) for them to make an informed choice. The appropriate remedies for each condition are obvious: improve the provision of disinterested (objective) information; use (disinterested) third-party validation against proven standards; develop better and more reliable labelling, information and communication procedures. However, as Keeling *et al.* (2012) find, governments are not always considered the most reliable or credible sources of disinterested information, or the most effective communicators, and there is a clear role for third parties (including NGOs) in improving information and communication about animal welfare conditions and concerns. In addition, 'consumers may use mistrust in information as an excuse of their unwillingness to change their purchasing behaviour in line with their alleged concerns' (Toma *et al.*, 2011, p. 263), which is, effectively, another version of the 'cheap talk' reason above. However, improved communications, information and more reliable animal welfare labels, while obvious and useful, may not necessarily narrow the gap between attitudes and actions substantially.

The fifth and sixth reasons (Fig. 2.2, 5, 6) for the disparity between reported attitudes/stated preferences and actual behaviour are probably more important (e.g. Verbeke, 2009). If consumers have to spend valuable time and energy searching for animal welfare-friendly products, then the effective price of improved animal welfare rises and the demand falls. Similarly, if other things (taste, quality, safety, convenience) are more important than the welfare provenance, then consumers will ration their time, effort and money in favour of these more important attributes and not bother to seek out specific animal welfare-friendly products. Moreover, consumers (and citizens) are not only concerned about their food; they have plenty of other things to be interested in, worried and concerned about, and on which to spend their time, effort and money. They will, in other words, tend to be 'rationally ignorant' – not bothering to spend time and effort trying to find out about, or to find, animal welfare-friendly products in the face of all their competing interests and (pre-)occupations. The benefits they would get from the effort are simply not worth it to them, though their responses in surveys may often be revealed as being 'unable to find welfare-friendly products' and is often ascribed to be due to a 'lack of availability'. It is also likely that at least some people would prefer not to be reminded about animal welfare at all when shopping for food. As a consequence, 'Improved farm animal welfare is more likely be realised and valued by consumers when it is integrated within a broader concept of quality, such as quality assurance or sustainability schemes' (Verbeke, 2009, p. 325). Both attitude surveys and typical WTP studies, being focused on the specific issue of animal welfare, in isolation from the pressures and distractions of the real world, do not include these last two real-world constraints and considerations. As a consequence, they are necessarily biased in favour of the focused issue. Isolated 'artificial' attitudes are not likely to be reliable indicators of purchase intentions or actions, as is frequently found in the literature, and as is well-recognized by marketers, who seldom use WTP estimates when trying to establish the likely market for new products.

2.5   Conclusion

Despite significant progress in animal welfare science, animal welfare remains controversial and subject to debate. Valuing the economic benefits society gains from better farm animal welfare is key if high(er) animal welfare standards are to be required and 'failure of viewing animal welfare in the context of other goals can ultimately have a negative effect on the animals' (Søresen et al., 2001). Their value and importance are intrinsically linked to their 'use value' or what they contribute to economic output. Nevertheless, society attaches an additional ethical value (a human construct) which undoubtedly cannot be ignored, particularly given the growing public concerns regarding the way farm animals are reared and the societal demand for governmental intervention (as expressed in public surveys).

A market system fails when the free market does not provide what society wants or it fails to achieve social efficiency. In the case of animal welfare, however, there is no sufficient (economic) evidence to indicate market failure, i.e. the achievement and delivery of socially optimal levels of animal welfare. The gaps between citizens' votes and consumer willingness to pay (and spend) do not necessarily indicate market failure. Additionally, animal welfare itself is not, strictly, a public good, though regulation of animal welfare is a public good; once it is provided for one person, it is also provided for everyone and no one can be prevented from enjoying the benefits of improved animal welfare. Animal welfare does suffer potential free-rider difficulties associated with consumption externalities but measurement of the extent of this problem, in practice, is fraught with difficulty. Only the free-rider problem, i.e. free-rider deficit bigger than market deficit, might indicate that legislation is required to solve 'market failure', and even then there are alternative remedies. Introducing legislation on the supposed grounds that the free-rider deficit is greater than the market deficit (the additional spending required to fund the desired improvement in animal welfare) runs the risk that the estimates of the deficits are wrong, so the legislation actually results in a decline in production; hence it can be argued that this is a case of 'government failure' rather than 'market failure'. In two of the six possible cases (presented in Fig. 2.2) where there is a citizen/consumer discrepancy, there is still no reason to invoke 'market failure', since in each case the market is actually telling us that the additional effort, time and resources necessary to bridge the gap between citizens' desires for better animal welfare and their willingness to pay for the provision of improved animal welfare are not worth the effort (the costs are not outweighed by the benefits). Markets allocate scarce resources among competing needs and desires, so they do not and cannot guarantee that each and every desire or want is actually met. We do know, however, that a shift in people's perceptions may be achieved once accurate and credible information about agricultural practices is provided. More honest and open communication between the industry and the public about food production could certainly be a helpful starting point. Farmers, no matter the system of production, need to be able to show consumers where their money goes, such as in ensuring good husbandry and high animal welfare. Shoppers should also be encouraged to consider the consequences of falling profit margins on producers and their animals. Researchers have their own role to play and we cannot continue to bandy about terms such as (good) animal welfare without pinning down the meaning more usefully. Until we do, the challenge to reach a proper understanding of how it may be improved, and how production systems can contribute, remains. If we could get to a point of common understanding of the term, we could at least begin to measure our degree of success or failure. Imposing higher standards than a society is willing to pay for, hence reducing freedom of choice, will lead to reduction in consumption levels or drive producers out of business. A market-driven approach, where any potential increase in production costs is compensated by premium prices under the prerequisite of willingness to pay by consumers (and retailers as intermediates), is preferable and more sustainable (Spoolder *et al.*, 2011). The food industry, in particular retailers, have been far more efficient at segmenting the market on animal welfare, bringing it to wider attention and providing a means for those with an interest in farm animal welfare to express their preferences (Buller *et al.*, 2018). Thus, it is the element of private (rather than public) good that works well, with markets allowing consumers to express their preferences through choices and purchases. As Norwood and Lusk (2011b) note: 'What people want is ultimately what matters. Even in the unlikely event that we could all agree that the goal is to maximize the well-being of farm animals, we humans are the ones that must decide how to make it happen.'

Notes

[1] Most of this section relies heavily on: https://www.staff.ncl.ac.uk/david.harvey/AEF873/PGandE/PGandE.html; (https://www.staff.ncl.ac.uk/david.harvey/AES829/Economics.html);and https://www.staff.ncl.ac.uk/david.harvey/ACE2006/MarketFailure/PGandE.html (accessed 17 April 2019).

[2] Concept first used by the economist Richard Musgrave in his paper 'A Multiple Theory of Budget Determination' (1957); however, the concept is highly controversial in economic theory.

[3] As above, section heavily based on: https://www.staff.ncl.ac.uk/david.harvey/ACE2006/MarketFailure/PGandE.html; and https://www.staff.ncl.ac.uk/david.harvey/AEF873/PGandE/PGandE.html; and https://www.staff.ncl.ac.uk/david.harvey/AES829/Economics.html(last accessed 17 April 2019).

[4] Available at: https://www.staff.ncl.ac.uk/david.harvey/ACE2006/MarketFailure/PGandE.html; and https://www.staff.ncl.ac.uk/david.harvey/AEF873/PGandE/PGandE.html (last accessed 17 April 2019).

[5] This is equal with the sum of each consumer's individual marginal benefit from that unit of good (Krugman and Welles, 2015, p. 499).

[6] https://www.staff.ncl.ac.uk/david.harvey/ACE2006/MarketFailure/PGandE.html; https://www.staff.ncl.ac.uk/david.harvey/AES829/Economics.html; https://www.staff.ncl.ac.uk/david.harvey/AEF873/Sustainability/Sust.html (last accessed 17 April 2019).

[7] This section reuses, extensively, material from Harvey and Hubbard (2013), which, in turn, stems from an intensive collaborative EU research project (EconWelfare) on the economics of animal welfare.

[8] Reused section from Harvey and Hubbard (2013).

[9] Again, we abstract from the considerable technical and practical difficulties of identifying exactly the spending specifically in favour of animal welfare (as opposed to other attributes of the products), and simply assume that we can do and have done this already.

[10] In technical terms, the ubiquitous assumption of independence of irrelevant alternatives, necessary for the estimation of WTP, is almost bound to be violated in practice. In effect, the assumption is analogous to the *ceteris paribus* assumption invoked so as to be able to employ partial equilibrium analysis of a complex general disequilibrium system. As Carson and Groves (2011) point out (p. 311), rational (and reasonable) responses to consequential questions practically guarantee 'non-truthful' answers, in the sense that responses are likely to deviate systematically from actual behaviour (the 'truth').

[11] We might also note that the consumption externalities which clearly exist for animal welfare also exist for many other goods and services – such as alcohol, pornography, energy-inefficient cars and so forth.

[12] Some might object that WTP and actual spending use both ability and willingness to pay as the appropriate weight to attach to individual preferences for animal welfare. To adopt any other weight, however, implies that we should treat animal welfare not as a *public* good, but as a *merit* good – to be supplied to all, free at the point of use, regardless of peoples' willingness or ability to pay. Extension of this argument implies some sort of deliberative social planning and management of animal production. The relative costs and benefits of this alternative are beyond the bounds of this paper, though the observation does have implications for the appropriate policy response.

References

Andersen, L.M. (2011) Animal Welfare and Eggs - Cheap Talk or Money on the Counter? *Journal of Agricultural Economics* 62(3), 565–584. DOI: 10.1111/j.1477-9552.2011.00310.x.

Baltzer, K. (2004) Consumers' willingness to pay for food quality – the case of eggs'. Section C – Food Economics. *Acta Agriculturae Scandinavica* 1(2), 78–90.

Bennett, R. and Blaney, R. (2002) Social consensus, moral intensity and willingness to pay to address a farm animal welfare issue. *Journal of Economic Psychology* 23(4), 501–520. DOI: 10.1016/S0167-4870(02)00098-3.

Blamey, R.K., Common, M.S. and Quiggin, J.C. (1995) Respondents to contingent valuation surveys: consumers or citizens? *Australian Journal of Agricultural Economics* 39(3), 263–288. DOI: 10.1111/j.1467-8489.1995.tb00554.x.

Buller, H. and Morris, C. (2003) Farm Animal Welfare: A New Repertoire of Nature-Society Relations or Modernism Re-embedded ? *Sociologia Ruralis* 43(3), 216–237. DOI: 10.1111/1467-9523.00242.

Buller, H., Blokhuis, H., Jensen, P., Keeling, L. and Harry, B. (2018) Towards farm animal welfare and sustainability. *Animals* 8(6), 81. DOI: 10.3390/ani8060081.

Carson, R.T. and Groves, T. (2007) Incentive and informational properties of preference questions. *Environmental and Resource Economics* 37(1), 181–210. DOI: 10.1007/s10640-007-9124-5.

Carson, R.T. and Groves, T. (2011) Incentive and information properties of preference questions: commentary and extensions. In: Bennett, J. (ed.) *International Handbook of Non-Market Environmental Valuation*. Edward Elgar, Northampton, Massachusetts.

Cicia, G. and Colantuoni, F. (2010) Willingness to pay for traceable meat attributes: a meta-analysis. *International Journal of Food System Dynamics* 3, 252–263.

Clark, B., Stewart, G.B., Panzone, L.A., Kyriazakis, I. and Frewer, L.J. (2016) A systematic review of public attitudes, perceptions and behaviours towards production diseases associated with farm animal welfare. *Journal of Agricultural and Environmental Ethics* 29(3), 455–478. DOI: 10.1007/s10806-016-9615-x.

Clark, B., Stewart, G.B., Panzone, L.A., Kyriazakis, I. and Frewer, L.J. (2017) Citizens, consumers and farm animal welfare: a meta-analysis of willingness-to-pay studies. *Food Policy* 68, 112–127. DOI: 10.1016/j.foodpol.2017.01.006.

Clark, B., Panzone, L.A., Stewart, G.B., Kyriazakis, I., Niemi, J.K. *et al.* (2019) Consumer attitudes towards production diseases in intensive production systems. *Plos One* 14(1), e0210432. DOI: 10.1371/journal.pone.0210432.

Cornish, A., Raubenheimer, D. and McGreevy, P. (2016) What we know about the public's level of concern for farm animal welfare in food production in developed countries. *Animals* 6(11), 74.

Curtis, J.A. and McConnell, K.E. (2002) The citizen versus consumer hypothesis: evidence from a contingent valuation survey. *The Australian Journal of Agricultural and Resource Economics* 46(1), 69–83. DOI: 10.1111/1467-8489.00167.

Diamond, P.A. and Hausman, J.A. (1994) Contingent valuation: is some number better than no number? *Journal of Economic Perspectives* 8(4), 45–64. DOI: 10.1257/jep.8.4.45.

Duncan, I.J.H. (2005) Science-based assessment of animal welfare: farm animals. *Revue Scientifique et Technique (International Office of Epizootics)* 24(2), 483–492.

Duncan, I.J.H. and Fraser, D. (1997) Understanding animal welfare. In: Appleby, M.C. and Hughes, B.O. (eds) *Animal Welfare*. CAB International, Wallingford, UK, pp. 19–31.

European Food Safety Authority (EFSA) (2015) *The use of animal-based measures to assess animal welfare in EU - state of the art of 10 years of activities and analysis of gaps*. Technical Report published 06 November 2015; EFSA Supporting publication 2015: EN-884. Available at: https://www.efsa.europa.eu/sites/default/files/scientific_output/files/main_documents/884e.pdf (accessed 5 October 2017).

European Union (1997) Treaty of Amsterdam amending the Treaty on European Union, the Treaties establishing the European communities and certain related acts. C 340. Official Journal of the European Communities, Luxembourg.

European Commission (2007) Eurobarometer: attitudes of EU citizens towards animal welfare, special Eurobarometer, 270/Wave 66.1, March. European Commission, Brussels.

European Commission (2016) Special Eurobarometer 442, summary, attitudes of Europeans towards animal welfare.

Fraser, D. (1999) Animal ethics and animal welfare science: bridging the two cultures. *Applied Animal Behaviour Science* 65(3), 171–189.

Fraser, D. and Weary, D.M. (2004) Quality of life of farm animals: linking science, ethics and animal welfare. In: Benson, G.J. and Rollin, B.R. (eds) *The Well-being of Farm Animals: Challenges and Solutions*. Blackwell, Ames, Iowa, pp. 39–60.

Graça, J., Calheiros, M.M. and Oliveira, A. (2016) Situating moral disengagement: motivated reasoning in meat consumption and substitution. *Personality and Individual Differences* 90, 353–364. DOI: 10.1016/j.paid.2015.11.042.

Hamilton, S.F., Sunding, D.L. and Zilberman, D. (2003) Public goods and the value of product quality regulations: the case of food safety. *Journal of Public Economics* 87(3–4), 799–817. DOI: 10.1016/S0047-2727(01)00103-7.

Harper, G. and Henson, S. (2001) *Consumer concerns about animal welfare and the impact on food choice*. Final report EU fair CT98-3678. Centre for Food Economics Research, University of Reading, UK.

Harvey, D. and Hubbard, C. (2013) Reconsidering the political economy of farm animal welfare: an anatomy of market failure. *Food Policy* 38, 105–114. DOI: 10.1016/j.foodpol.2012.11.006.

Hewson, C. (2003) What is animal welfare? Common definitions and their practical consequences. *The Canadian Veterinary Journal* 44, 496–499.

Hubbard, C., Bourlakis, M. and Garrod, G. (2007) Pig in the middle: farmers and the delivery of farm animal welfare standards. *British Food Journal* 109(11), 919–930. DOI: 10.1108/00070700710835723.

Hubbard, C., Clark, B. and Foster, L. (2018) Farm animal welfare and consumers. In: Matsumoto, S. and Otsuki, T. (eds) *Consumer Perception of Food Attributes*. CRC Press, Taylor & Francis, pp. 171–187.

Hubbard, C. and Scott, K. (2011) Do farmers and scientists differ in their understanding and assessment of farm animal welfare? *Animal Welfare* 20(1), 79–87.

Keeling, L.J., Immink, V., Hubbard, C., Garrod, G., Edwards, S.A. *et al.* (2012) Designing animal welfare policies and monitoring progress. *Animal Welfare* 21(1), 95–105. DOI: 10.7120/096272812X13345905673845.

Kjærnes, U. (2012) Ethics and action: a relational perspective on consumer choice in the European politics of food. *Journal of Agricultural and Environmental Ethics* 25(2), 145–162. DOI: 10.1007/s10806-011-9315-5.

Kjaernes, U. and Lavik, R. (2007) *Farm Animal Welfare and Food Consumption Practices: Results from Surveys in Seven Countries*. School of City and Regional Planning, Cardiff University, Cardiff, UK.

Krugman, P. and Welles, R. (2015) *Economics*, 4th edn. Worth Publishers, New York.

Lagerkvist, C.J. and Hess, S. (2011) A meta-analysis of consumer willingness to pay for farm animal welfare. *European Review of Agricultural Economics* 38(1), 55–78. DOI: 10.1093/erae/jbq043.

Loughnan, S., Bastian, B. and Haslam, N. (2014) The psychology of eating animals. *Current Directions in Psychological Science* 23(2), 104–108. DOI: 10.1177/0963721414525781.

Lusk, J.L. (2011) The market for animal welfare. *Agriculture and Human Values* 28(4), 561–575. DOI: 10.1007/s10460-011-9318-x.

Lusk, J.L., Roosen, J. and Fox, J.A. (2003) Demand for beef from cattle administered growth hormones or fed genetically modified corn: a comparison of consumers in France, Germany, the United Kingdom, and the United States. *American Journal of Agricultural Economics* 85(1), 16–29. DOI: 10.1111/1467-8276.00100.

Mann, S. (2005) Ethological farm programs and the "market" for animal welfare. *Journal of Agricultural and Environmental Ethics* 18(4), 369–382. DOI: 10.1007/s10806-005-7049-y.

McInerney, J.P. (2004a) *Animal welfare, Economics and Policy, Report on a study undertaken for the Farm & Animal Health Economics Division of Defra*. UK.

McInerney, J.P. (2004b) Animal welfare, economics and policy. *Journal of the Royal Agricultural Society of England* 165, ISSN, 0800–4134.

McVittie, A., Moran, D., Sandilands, V. and Spaks, N. (2006) *Estimating non-market benefits of reduced stocking density and other welfare increasing measures for meat chickens in England*, final report to Defra AW0236.

Norwood, B.F. and Lusk, J.L. (2011a) *Animal welfare economics. applied economic perspectives and policy*, 1-21, advanced access published November 17.

Norwood, B.F. and Lusk, J.L. (2011b) *Compassion by the Pound: The Economics of Farm Animal Welfare*. Oxford University Press, Oxford, UK.

Rothgerber, H. and Rothberger, H. (2014) Efforts to overcome vegetarian-induced dissonance among meat eaters. *Appetite* 79, 32–41. DOI: 10.1016/j.appet.2014.04.003.

Sandøe, P., Christiansen, S.B. and Appleby, M.C. (2003) Farm animal welfare: the interaction of ethical questions and animal welfare science. *Animal welfare* 12, 469–478.

Schmid, O. and Kilchsperger, R. (2011) Farm Animal Welfare legislation and standards in Europe and world-wide: a comparison with the EU regulatory framework. *Proceedings of the Third Scientific Conference of ISOFAR 'Organic is Life: Knowledge for Tomorrow'*, Manyangju, Korea, 26 September-1 October, pp. 104–107.

Scott, E.M., Nolan, A.M. and Fitzpatrick, J.L. (2001) Conceptual and methodological issues related to welfare assessment: a framework for measurement. *Acta Agriculturae Scandinavica, Section A — Animal Science* 51(sup030), 5–10. DOI: 10.1080/090647001316922983.

Sloman, J. (2014) *Essential Economics for Business*, 4th edn. Pearson Education Limited, Harlow, UK.

Sloman, J. and Garrat, D. (2017) *Essentials of Economics*, 7th edn. Pearson Education Limited, Harlow, UK.

Spoolder, H., Bokma, M., Harvey, D., Keeling, L., Majewsky, E. *et al.* (2011) EconWelfare findings, conclusions and recommendations concerning effective policy instruments in the route towards higher animal welfare in the EU; deliverable 5.0.

Spooner, J.M., Schuppli, C.A. and Fraser, D. (2014) Attitudes of Canadian citizens toward farm animal welfare: a qualitative study. *Livestock Science* 163, 150–158. DOI: 10.1016/j.livsci.2014.02.011.

Søresen, J.T., Sandøe, P. and Halberg, N. (2001) Animal welfare as one among several values to be considered at farm level: the idea of an ethical account for livestock farming. *Acta Agriculturae Scandinavica* 30(Suppl), 11–16.

Toma, L., McVittie, A., Hubbard, C. and Stott, A.W. (2011) A structural equation model of the factors influencing British consumers' behaviour toward animal welfare. *Journal of Food Products Marketing* 17(2–3), 261–278. DOI: 10.1080/10454446.2011.548748.

Uehleke, R. and Hüttel, S. (2018) The free-rider deficit in the demand for farm animal welfare-labelled meat. *European Review of Agricultural Economics* 46(2), 291–318. DOI: 10.1093/erae/jby025.

Vanhonacker, F., Verbeke, W., van Poucke, E. and Tuyttens, F.A.M. (2007) Segmentation based on consumers' perceived importance and attitude toward farm animal welfare. *International Journal of Sociology of Food and Agriculture* 15(3), 84–100.

Verbeke, W. (2009) Stakeholder, citizen and consumer interests in farm animal welfare. *Animal Welfare* 18, 325–333.

YouGov (2019) Is the future of food flexitarian?

3 Consumer Demand for Animal Welfare Products

FAICAL AKAICHI,* AND CESAR REVOREDO-GIHA

Food Marketing Research, Rural Economy, Environment and Society Department, Scotland's Rural College (SRUC)

Summary

Modern agricultural practices have increased the efficiency of food production with a decrease in their cost and prices for consumers. However, to some extent this has been detrimental to the ethical way in which livestock are treated, particularly in more intensive production systems. On the demand side, an increasing number of consumers are interested in the way that food is produced and the attributes behind it. Animal welfare is one of those ethical attributes that are particularly important for consumers, and at the retail level, it is reflected in a number of labels aiming at passing cues (due to its nature as a credence attribute) to consumers. For meat supply chains, these labels have the possibility to positively affect sales if consumers are willing to pay more for products with those attributes. Moreover, if increasing animal welfare implies higher costs of production, it is important for the supply chain to know whether these costs can be passed on to consumers. These issues have motivated a substantive literature on the measurement of consumers' interest in animal welfare and their willingness to pay for its attributes. The purpose of this chapter is to provide an overview of the economic theory behind the measurement of animal welfare and some empirical applications.

3.1 Introduction

The way animals are treated has also been affected by the interacting forces of rising demand for cheaper meat products, increasing competition in the global market and a shift to mechanization and automation in agriculture, which have encouraged livestock farmers to adopt intensive livestock

---

*Corresponding author: Faical.Akaichi@sruc.ac.uk

systems (e.g. indoor housing systems) to maximize their production given limited space and resources (Blount, 1968; Cronin *et al.*, 2014).

Consequently, farmers (and further down the supply chain) have come under increasing pressure to address public concern about negative externalities of intensive livestock systems, especially in terms of animal welfare (e.g. confinement and related health problems, tail docking, beak trimming, castration, mixing aggression) and environmental effects (e.g. soil and underground water pollution) (Norwood and Lusk, 2011a). In fact, animal advocacy groups have pressured policy makers to outlaw certain production practices (e.g. battery cages, gestation crates) and force farmers to use alternative production systems that are perceived to have less negative effects on animal welfare. The European Union recognized that animals are sentient beings and specified the minimum standards that ensure that they do not endure avoidable pain or suffering during the production process, during transport or at slaughter. For instance, in January 2006, the Commission adopted a Community Action Plan for the protection and welfare of animals for the period 2006–2010 (European Commission, 2007). Furthermore, farmers who voluntarily adopted animal-friendly production standards were able to have their products labelled as animal-friendly to inform consumers that they were purchasing produce of high animal welfare standards.

As regards consumers' interest in animal welfare, as pointed out by Verbeke (2009), the Eurobarometer survey carried out in 2005 showed that European citizens were concerned about the welfare of farm animals. The survey indicated that 82.3% of Europeans evaluated the overall welfare of farm animals as being somewhere between moderate and very bad, with 78.3% strongly believing that more should be done to improve the welfare and protect the living conditions of farm animals in the EU. Additionally, almost 90% of respondents claimed not to receive sufficient information about the welfare and protection of farm animals. Moreover, animal welfare was identified as one of the key areas on animals and the environment of relevance to future patterns of livestock and meat production (Grunert, 2006).

Animal welfare is one of those ethical attributes that are particularly important for consumers, and at the retail level, it is reflected in a number of labels aiming at passing cues (due to its nature as a credence attribute) to consumers. For meat supply chains, these labels have the possibility to positively affect their sales if consumers are willing to pay more for products with those attributes. Moreover, if increasing animal welfare implies higher costs of production, it is important for the supply chain to know whether these costs can be passed to consumers.

The above issues have motivated a substantive literature on the measurement of consumers' interest in animal welfare and their willingness to pay for its attributes. The purpose of this chapter is to provide an overview of work on animal welfare from the demand side.

The chapter starts with a review of the economic theory behind the measurement of consumer demand for products with animal welfare attributes and the willingness to pay for these attributes. It is followed by a literature review of empirical applications and two case studies.

3.2 Theory and Empirical Methods

3.2.1 Demand theory

Conventional demand theory is largely centred on the fact that consumers are rational and, hence, will carefully evaluate and consume the bundle of products that maximize the utility generated from the satisfaction of their needs and desires. In traditional demand theory, consumer utility is expressed by:

$$U = f\left(C_1, C_2 \ldots C_K\right)$$

Where the utility (U) is a function of different commodities facing consumers, and over which they are expected to choose the commodity (or bundle of commodities) that maximizes their utility. Demand theory proposes that although it is impossible to measure the utility that the consumer derives from buying and consuming a commodity, it is possible to rank the commodities in their order of preference to the consumer. This implies that it is possible to assess, among others, whether a product with high animal welfare is more or less preferred than other products in the same set of commodities. In addition, the ratio $\partial U / \partial C\_k$ is indicative of the price that consumers are willing to pay to purchase the commodity C_k.

A key principle of conventional demand theory is that the maximization of consumer utility is constrained by a consumer's income and the prices of the commodities. This implies that a consumer's decision about how much to buy of a commodity is affected, among other things, by the price of the product of interest, the prices of the other products in the consumption sets and his/her income. In demand theory, the demand for a product is typically represented as:

$$Q_j = f\left(P_j, P_k, I, Z\right)$$

Where Q_j is the quantity demanded of product j which is a function of the price of product j (P_j), the prices of the products in the consumption set (P_k), consumer's income I, and other factors Z, such as consumers' preferences and socio-demographic factors.

In empirical demand analysis that aims to measure the sensitivity of demand to changes in a product's price and consumer income, economists usually use a system of demand equations to derive a key indicator of demand sensitivity to changes in price and income, termed 'elasticity'. There are three types of elasticities: own-price, cross-price and income elasticity. Own-price elasticity is a measure of the percentage change in the quantity demanded of product J as a result of 1% change in the price of product J. Cross-price elasticity is a measure of the percentage change in the quantity demanded of product A as a result of 1% change in the price of anther product (say product I). The income (or expenditure) elasticity is a measure of the percentage change in the quantity demanded of a product 'caused' by a 1% change in consumers' income. These elasticities are computed based on the output obtained from the estimation of a demand system.

Perhaps the most widely used system of demand equations (known as a demand system) to analyse the demand for food products in general, and animal-friendly food products in particular, is the linear version of the Almost Ideal Demand System (AIDS).

The AIDS model was pioneered by Deaton and Muellbauer (1980). The popularity of this demand model among demand analysts is due to its flexibility to include parametric restrictions required for consistency with economic theory (Deaton and Muellbauer, 1980).

The AIDS model is generated from a cost-minimization problem that defines the minimum expenditure necessary for a consumer to attain a specific level of utility at a given set of prices. The demand functions are obtained in the share of consumer's budget spent on product i, in time t (i.e. w_{it}). The budget shares are obtained by logarithmic differentiation of the expenditure function with respect to prices. These shares are given by:

$$w_{it} = a_i + \sum_{j=1}^{n} \gamma_{ij} ln p_{jt} + \beta_i ln \left(\frac{x_t}{p_t} \right)$$

where the shares are a function of the price of commodity j (p_{jt}) and the total expenditure x_t. The price index p_t is used as a deflator to express the total expenditure in real terms. a_i is the constant coefficient (i.e. intercept) in the ith share equation; γ_{ij} is the slope coefficient associated with the jth good in the ith share equation. It represents the change in the ith product's budget share with respect to a change in jth price with real expenditure held constant. The coefficient β_i represents the change in the ith product's budget share with respect to a change in real expenditures with price held constant. The analysis consists in estimating the parameters γ_{ij} and β_i which will be then used to compute the conditional own- (e_{ii}) and cross-price (e_{ij}) elasticities as well as the expenditure elasticity (e_i) as follows[1].

$$e_{ii} = \frac{\gamma_{ij}}{w_i} - \frac{\beta_i}{w_i} \left(\alpha_j + \sum_{k=1}^{n} \gamma_{ij} ln p_k \right) - 1$$

$$e_{ij} = \frac{\gamma_{ij}}{w_i} - \frac{\beta_i}{w_i} \left(\alpha_j + \sum_{k=1}^{n} \gamma_{ij} ln p_k \right)$$

$$e_i = 1 + \frac{\beta_i}{w_i}$$

Few research studies have assessed consumer demand for animal-friendly food products using market data and estimating demand system models (e.g. Lusk, 2010; Tonsor and Olynk, 2011). This is probably due to the relatively recent use of labels to certify food products with animal welfare attributes. Consequently, the purchases of animal-friendly food products has not been explicitly distinguished from other food products in market data until recently.

The traditional demand theory posits that consumers have preferences over goods and that demand for a commodity can be altered only if its price or/and consumer's income is (are) altered too. This implies that there is no way the effect of changes in a product's non-price attributes, such as quality and labelling, can be measured and analysed if those changes do not alter the price of the product.

Lancaster (1966) proposed a new and conceptually more realistic framework for explaining consumer demand. His approach is based on the principle that the consumers demand products for their characteristics or attributes (e.g. taste, origin, nutritional content, level of convenience, ethical attributes, price), not the product itself. This approach has the advantage of allowing demand analysts to predict how demand will be affected by a change in one or more attributes (e.g. change in the level of animal welfare) as well as to assess whether consumers trade off the attributes when deciding which product in the consumption set to buy.

In the Lancaster model of consumer demand, utility functions are defined not in terms of products consumed but by the characteristics (X_k) that give value in consumption:

$$U = f\left(X_1, X_2 \ldots X_K\right)$$

Several research studies (e.g. Lagerkvist *et al.*, 2006; Liljenstolpe, 2008; Gracia *et al.*, 2014) on consumer demand for food products with high animal welfare have used the Lancaster model framework. In those studies, the studied food products have been described to consumers as a combination of different attributes (e.g. animal welfare, organic, local, fat content and price). In particular, consumers were presented with several sets of alternatives. Each alternative is a combination of food attributes, including the attribute animal welfare.

The consumer's task in these studies consists of comparing the combinations in each set and indicating the combination he/she prefers most. The most commonly used choice-elicitation method is the discrete choice experiment. In a discrete choice experiment, participants are generally first asked to participate in a choice task; then they are required to complete a questionnaire about their attitudes toward animal welfare and related issues as well as their socio-demographic characteristics. In the choice task, participants are successively provided with a few different choice sets and are repeatedly asked to choose between at least two different alternatives (see an example of a choice set in Fig. 3.1).

Several choice models have been devised to analyse choice data obtained using discrete choice experiments. These models allow the choice analyst to gain information on consumers' preferences and willingness to pay for each attribute of the studied product as well as whether and how they trade off the product's attributes considered in the study. The choice data are generally analysed within a random utility framework (McFadden, 1974). Thus an individual n presented with j alternatives at a choice occasion (set) t is expected to choose the alternative that maximizes his/her utility.

Following Lancaster's concept that any product is a bundle of attributes, the utility that individuals derive from the consumption of a product is assumed to be equal to the sum of their marginal utility for each of the attributes that constitute the product of interest. Consequently, if we assume a sample of N respondents who are presented with T choice occasions of J alternatives each, individual n's utility (U_{njt}) from choosing the jth alternative at a tth choice occasion takes the form:

Choice set 3.2			Identification number:		
Attributes	**OPTION 1**	**OPTION 2**	**OPTION 3**	**OPTION 4**	**OPTION 5**
Animal welfare:	Animal friendly	No label	Animal friendly	No label	
Organic:	Not organic	Not organic	Organic	Organic	
Local:	Not local	Local	Local	Not local	None of them
Fat content:	Low fat	No label	No label	Low fat	
Price:	£4.49	£3.79	£5.29	£4.49	
	□	□	□	□	□
Please mark the option you would choose.					

Fig. 3.1. An example of a choice set used in the choice experiment.

$$U_{njt} = V_{njt} + \varepsilon_{njt}$$

where V_{njt} is the deterministic component and ε_{njt} is the random compo-nent. ε_{njt} is assumed to be independent and identically distributed. Assuming that the deterministic component of the utility is linear-in-parameter, V_{njt} can be written as:

$$U_{njt} = \beta X_{njt} + \varepsilon_{njt}$$

where β denotes the $K \times 1$ vector of unknown utility parameters and X_{njt} represents the attributes' levels considered in the study (e.g. high animal welfare, low fat content, organic, low greenhouse gas emissions).

The conditional logit (CL) model (McFadden, 1974) is the workhorse choice model for analysing discrete choice data. However, its assumptions (i.e. homogeneity of consumers' preferences and the alternatives included in any choice set are treated by respondents as independent) were found to be unrealistic and do not generally hold (Hensher *et al.*, 2015). Train (1998) proposed a less restrictive model (Random Parameter Logit (RPL)) that allows individuals' preferences to be heterogeneous and the assumption of the Independence of Irrelevant Alternatives to be relaxed. In the RPL, at least some of the parameters are specified as random. In other words, each individual is considered to have a unique set of preferences, reflected in the individual parameters β_i. In the RPL, the conditional choice prob-ability that individual i chooses an alternative j at a choice occasion t is specified as:

$$P\left(j|X_{it},\beta\right) = \prod_{t=1}^{T}\left[\frac{\exp\left(\beta_i' X_{ijt}\right)}{\sum_{k=1}^{J}\exp\left(\beta_i' X_{ikt}\right)}\right]$$

The unconditional choice probability is the expected value of the logit probability integrated over all possible values of β and weighted by the density of β:

$$\int_{\beta} P\left(j|X_{it},\beta\right)f\left(\beta|\Omega\right)d\beta$$

Since $P(j\,|\,X_{it},\Omega)$ does not have a closed form solution, it is therefore approximated through simulation methods. In particular, R draws of β_{ir} are taken from the distribution $f(\beta\,|\,\Omega)$. For each draw, the choice probability is calculated. Then the resulting probabilities from the R draws are averaged. The simulated log-likelihood (SLL) for all respondents, which is estimated via maximum likelihood procedures, is calculated as:

$$SLL = \sum_{i=1}^{I}\sum_{t=1}^{T}ln\left(\frac{1}{R}\sum_{r=1}^{R}\frac{\exp\left(\beta_{ir}X_{ijt}\right)}{\sum_{k=1}^{J}\exp\left(\beta_{ir}X_{ikt}\right)}\right)$$

In addition to the estimation of respondents' preferences, choice data are often used to calculate consumers' willingness to pay (WTP). WTP is commonly expressed as the negative ratio of the non-price attribute coefficient to the price attribute coefficient.

$$WTP_{non-price\ attribute} = -\frac{\beta_{non\ price\ attribute}}{\beta_{price}}$$

3.3 Empirical Results

3.3.1 Literature review

Understanding consumers' purchasing behaviour as well as its determining factors (e.g. consumers' attitudes and socio-demographic characteristics) is key to boosting the demand for animal-friendly food products through developing products and marketing strategies tailored to consumers' needs.

There is an increasing body of literature on consumers' demand for food product with high animal welfare and its determinants. In this section, we will summarize the findings from previous research studies on two main topics: (i) consumers' attitudes towards farm animal welfare and related issues such as barriers to the demand for animal-friendly products and the profile of animal welfare-minded consumers; and (ii) consumers' demand and WTP for animal welfare.

Several studies have investigated consumers' attitudes towards farm animal welfare and have showed an agreement among citizens in developed countries on the principle that farmed animals should be treated humanely and that cruelty towards them is unacceptable (e.g. Harper and Henson, 2001; Schroder and McEachern, 2004; Mayfield *et al.*, 2007; Honkanen and Ottar Olsen, 2009; Norwood and Lusk, 2011b; Kehlbacher *et al.*, 2012). For example, a study carried out in 25 European countries as part of a research project funded by the European Commission found that animal welfare was highly recognized by the European citizens who attributed, on a scale from 1 to 10, an average of 7.8 to the importance of protecting farmed animals. As expected, the attitudes of respondents across Europe were found to be heterogeneous. The importance given to animal welfare was found to be higher in Scandinavian countries (e.g. Sweden (9.0), Finland (8.7) and Denmark (8.6)) and lower in other countries such as Lithuania (6.9) and Spain (6.9) (European Commission, 2007). Heterogeneity of consumers' attitudes across animal species was also documented. For instance, consumers' concerns about the way animals are farmed were found to be higher for hens and pigs (Lagerkvist *et al.*, 2006) and lower for farmed fish (Honkanen and Ottar Olsen, 2009). Despite consumers showing high interest in animal welfare, research has also revealed that the attribute animal welfare is not the main determinant when deciding which food product to buy. For instance, consumers mentioned price, quality, health benefits and product safety as more important food attributes (Kanis *et al.*, 2003; Lo and Matthews, 2002; Hutchins and Greenhalgh, 1997; Verbeke *et al.*, 1999; Lusk *et al.*, 2007; Vanhonacker *et al.*, 2010; Toma *et al.*, 2011).

Health, quality and food safety were mentioned by EU citizens as the main drivers of their demand for animal-friendly food products (European Commission, 2007). For example, in the UK, consumers revealed that they mainly buy animal-friendly meat because they think it is healthier (78%), safer (75%), better for the environment (72%), more nutritional (72%) and tastier (72%) (Kehlbacher *et al.*, 2012). The main barriers to the purchase of animal-friendly food products stated by consumers include limited availability of animal-friendly products including ready meals (e.g. in restaurants and small- to medium-sized grocery stores), lack of information on livestock husbandry, no labelling or inadequate labelling of animal-friendly food products (e.g. information on the back of the package or messages that are too technical or inexplicit (freedom food)), low trust in a certification scheme, consumers' lack of belief in their ability to improve animal welfare through their purchases, the higher retail prices of animal-friendly products, and lack of time to look for and read food labels (Harper and Henson, 2001; Schroder and McEachern, 2004; European Commission, 2007; Nocella *et al.*, 2010; Toma *et al.*, 2010; Vanhonacker *et al.*, 2010; Norwood and Lusk, 2011a; Lagerkvist and Hess, 2011).

When asked about which stakeholders they believe can best ensure that farm animals destined for human consumption are raised to high animal welfare standards, animal welfare-minded EU consumers reported that farmers and veterinarians are the most responsible. They also believe that governments should keep playing the role of regulator, while animal welfare activists should keep pressuring governments to improve farm animal welfare (European Commission, 2007). Interestingly, in the USA, Lusk and Norwood (2008) found that the majority of consumers believe that livestock husbandry standards should be decided by experts, rather than being based on public opinion (i.e. referendum). They also reported that decisions about how farm animals should be treated have to be based on scientific measures of animal wellbeing, rather than on moral and ethical considerations.

To help the development of policy strategies tailored to the needs of ethically minded consumers and to target them efficiently, some studies have looked at the consumers who were found to be more likely to buy and consume animal-friendly food products. These consumers have been described as female adults from more affluent households with highly educated members (Harper and Henson, 2001; Hill and Lynchehaun, 2002; Lagerkvist et al., 2006; Toma et al., 2010). Saying this, it is noteworthy that in addition to consumers' socio-demographic and economic characteristics, their preferences and WTP for farm animal welfare were also found to be determined by their attitudes toward ethical food attributes as well as their lifestyle (Norwood and Lusk, 2011a; Kehlbacher et al., 2012).

Measuring consumers' WTP for animal-friendly food products has been the main focus of many research studies (Bennett, 1997; European Commission, 2005; Lusk et al., 2006; Lagerkvist et al., 2006; Carlsson et al., 2007a, Carlsson et al., 2007b; Liljenstolpe, 2008; Tonsor et al., 2009; Raun Mørkbak et al., 2010; Norwood and Lusk, 2011a; Lagerkvist and Hess, 2011; Gracia et al., 2011; Kehlbacher et al., 2012). These studies revealed a strong correlation between consumers' attitudes and their WTP for food products with a high level of animal welfare. In other words, consumers who were revealed to be concerned about how animals are farmed also reported a high price premium for animal-friendly food products. This price premium was found to vary across products, countries and even across data collection methods.

For instance, the results from a study that was carried out for the European Commission in 25 European countries found that, on average, 57% of consumers in Europe were willing to pay a price premium for animal welfare-friendly food products of at least 5%. Interestingly, the percentage of consumers who were found to be unwilling to pay a price premium for food products labelled as animal-friendly was the lowest (less than 30%) in the Scandinavian countries, the UK and The Netherlands, and the highest in Hungary and Slovakia (57%) (European Commission, 2005). In the USA, consumers were also revealed to be willing to pay a price premium for animal-friendly foods (Tonsor et al., 2009; Lusk et al., 2006; Norwood and Lusk, 2011a). For example, Norwood and Lusk (2011a) conducted a non-hypothetical experimental auction and found that US consumers are willing to pay up to 141% and 112% more for animal-friendly eggs and pork, respectively.

Abundant research in the literature on animal welfare has examined consumers' preferences and willingness to pay for animal-friendly food products. However, very little work has been devoted to assessing whether consumers treat animal welfare and other ethical food attributes as related and, if so, whether they consider them as substitutes (overlapping) or complements (Gracia *et al.*, 2014). Assessing the interactions between these attributes is of great importance because firms engaged in producing and marketing animal-friendly food products are also interested in carefully evaluating whether consumers trade off the attribute animal welfare with other desirable food attributes (e.g. local, organic) when purchasing food products. To fill this gap we carried a choice experiment to look at consumers' preferences and willingness to pay for four pork attributes (i.e. animal welfare, organic, local and low fat content) as well as whether and how they trade off these attributes. The results are presented below in the first empirical study.

As mentioned, most of the research studies that have investigated consumers' preferences and WTP for animal-friendly food products used self-reported surveys and hypothetical economic experiments. The advantages of these methods are that they are generally less expensive, easy to implement and offer researchers more flexibility to investigate novel attributes and claims as well as getting more attitudinal and sensory information compared with market data. However, participants in self-reported surveys and hypothetical economic experiments (e.g. contingent valuation and choice experiment) were found to overstate their WTP (i.e. hypothetical bias) due to absence of a monetary cost that forces/incentivizes consumers to reveal their true preferences and WTP.

Fortunately, if the study's main objective is to investigate consumer demand, the effect of hypothetical bias could be ruled out using data on actual consumers' purchases from grocery stores (e.g. scanner data). Only a few peer-reviewed studies (Baltzer, 2004; Lusk, 2010; Chang *et al.*, 2010; Tonsor and Olynk, 2011) used market data to explore the demand for animal-friendly food products. To further enrich this literature with more evidence, we used market data on food and drink purchases for consumption at home to measure consumers' demand for three types of pork products (i.e. conventional, animal-friendly and organic) as well as investigate whether their demand varies with the level of deprivation of the household. The results are presented in the second empirical study below.

3.3.2   Empirical study 1 – consumer choice experiment for different attributes of pork products

This study consisted of a choice experiment to assess consumers' preferences and WTP for four claims: 'Animal Friendly', 'Organic', 'Local' and 'Low Fat', as well as how consumers of pork trade off these claims. In addition, two consumer segments were also investigated to assess their differences, i.e. supporters of governmental intervention to regulate farm animal welfare and opponents of this type of intervention.

To collect the choice data, a laboratory choice experiment was conducted in Scotland. In total, 120 consumers were recruited from the city of Edinburgh and its metropolitan area by a market research company. Each consumer participated in only one of the 12 sessions of the choice experiment, of approximately one hour, and was paid a £35 participation fee. The company was asked to recruit a sample that was representative with respect to key socio-demographic characteristics of UK shoppers of food products. In the choice experiment, participants were first asked to participate in a choice task. Then they were required to complete a questionnaire about their attitudes towards animal welfare and related issues as well as their socio-demographic characteristics.

In the choice task, participants were successively provided with eight different choice sets and were repeatedly asked to choose between four different alternatives of fresh pork (300 g) and an opt-out alternative. Each alternative of fresh pork was a combination of different levels of five attributes: animal welfare (two levels: Animal friendly/No label), organic (two levels: Organic/Not organic), locality of the product (two levels: Local/Not local), fat content (two levels: Low fat content/No label) and the price (four levels: £3.19, £3.79, £4.49, £5.29). The choice experiment was designed following the approach proposed by Street and Burgess (2007). A fractional factorial design that is statistically efficient and allows the estimation of all of the main and the two-way interaction effects was generated. An example of one of the choice sets that was used in the choice experiment is displayed in Fig. 3.1.

To analyse the choice data, the random parameter logit model was used. The mean and the standard deviation of six main effects and six interaction effects were estimated. The deterministic component of the utility is expressed as follows:

$$
\begin{aligned}
\beta X_{njt} = {} & \beta_{price}Price + \beta_{None}None + \beta_{AF}Animal\ Friendly + \beta_{Org}Organic + \beta_{Loc}Local \\
& + \beta_{LF}Low\ Fat + \beta_{AFOrg}Animal\ Friendly * Organic + \beta_{AFLoc}Animal \\
& Friendly * Local + \beta_{AFLF}Animal\ Friendly * Low\ Fat + \beta_{OrgLoc}Organic * \\
& Local + \beta_{OrgLF}Organic * Low\ Fat + \beta_{LocLF}Local * Low\ Fat
\end{aligned}
$$

In the estimated random parameter logit model, the parameters for all the non-price attributes as well as the six two-way interactions were assumed to be normally distributed. For the price coefficient, a log-normal distribution was imposed to avoid obtaining unrealistic positive values. The random parameter logit model was estimated in WTP space following Train and Weeks (2005). The results are displayed in Table 3.1.

The results show that, on average, the sampled consumers are willing to pay significant price premiums equal to £1.94, £0.75, £1.38, and £1.31 for the claims 'Animal Friendly', 'Organic', 'Local' and 'Low Fat', respectively. Interestingly, the price premiums for the claims 'Animal Friendly', 'Local' and 'Low Fat' (Model 1) are significantly higher than their current retail price premium; however, the price premium for the claim 'Organic' is substantially lower that its actual retail price. The results were in line with findings from a large number of previous studies in that (i) there is a potential market for food products labelled as animal friendly, organic, local or healthier; and (ii) the high current retail price of organic foods is a major barrier to the translation of consumers' positive preferences for these types of foods into actual purchases.

Table 3.1. Estimated consumers' willingness to pay (in £ per 500 g of fresh pork).

Variables	Estimated mean	Estimated standard deviations	% respondents with positive WTP
Random parameters			
Animal Friendly	1.94***	1.16***	95%
Organic	0.75***	−0.06	100%
Local	1.38***	−0.05***	100%
Low Fat	1.31***	−1.07***	89%
Animal Friendly *Organic	−0.15***	−0.43***	36%
Animal Friendly *Local	−0.29***	0.02	0%
Animal Friendly *Low Fat	−0.27***	−0.41***	26%
Organic*Local	0.02	−0.16***	50%
Organic*Low Fat	0.04	0.06***	50%
Local*Low Fat	−0.33***	−0.31***	14%
Number of observations	960		
Total number of estimated parameters	69		
Log-likelihood (constant only)	−1514.686		
Final log-likelihood	−773.308		
McFadden's Pseudo R2	0.44		

(***) (**) and (*) denote statistical significance at (1%), (5%) and (10%) level, respectively.

Regarding the interaction effects, the results suggest that consumers do not actually treat all the desirable food attributes as independent when bundled. Only two out of the six estimated interaction effects were insignificant (i.e. independent attributes). In particular, it was found that, on average, participants in the study treated the attribute bundles 'Animal Friendly & Organic', 'Animal Friendly & Local', 'Animal Friendly & Low Fat', and 'Local Low Fat' as partially overlapping (i.e. substitutes), which in turn resulted in a significant discount of their price premium for each of these bundles. As expected, the amount of the discount (i.e. negative interaction) is significantly smaller than the main effect. This suggests that marketers can still bundle these claims, but before they do they need to make sure the benefit from bundling the attributes is higher than the discount effect. In addition to the policy, manufacturing and marketing implications of these results, they also suggest that total WTP estimates may be biased if complementarity and substitution effects are ignored when specifying and estimating the choice model.

The statistical significance of the estimated standard deviation suggests consumers' preferences and WTP for the four claims as well as most of the estimated interactions are heterogeneous among consumers. Since the non-price parameters were assumed to be normally distributed, the Z-test approach proposed by Train (2003) was used to partially explore the heterogeneity of respondents' WTP. This approach consists in computing the percentage of respondents who placed a positive (or negative) value on the non-price attributes as well as their interactions using the following formula:

$$100 * \Phi \left(\frac{-\beta_k}{S_k} \right)$$

where Φ is the cumulative standard normal distribution and β_k and S_k are the mean and the standard deviation of the *kth* coefficient, respectively.

Results presented in the last column of Table 3.1 show that most respondents positively valued the labels 'Animal Friendly', 'Organic', 'Local' and 'Low Fat'. None the less, their price premiums for bundles of these labels are quite heterogeneous. For example, the results show that while the majority of respondents perceived the labels 'Animal Friendly' and 'Organic' ('Low Fat') as substitutes, 36% (26%) of them perceived the two labels as complements. More interestingly, the non-significance of the estimated marginal WTP for the bundle 'Organic*Local' suggests that respondents perceived the labels 'Organic' and 'Local' as independent. None the less, the results from the Z-test computation show that half of respondents considered the two labels as complements while the other half perceived them as substitutes. As a result, the two effects cancelled out each other leading to a non-significant estimated marginal WTP for the bundle. The same also applies to the interaction 'Organic*Low Fat'. The results from the Z-test values show the importance of (i) estimating the interactions between desirable food attributes; and (ii) taking into account the heterogeneity of consumers' preferences and WTP (e.g. through segmenting consumers) to effectively increase consumer demand for desirable food attributes.

Finally, it is important to highlight that the data collected in this study cannot be used to investigate which approach (i.e. legislation versus labelling) should be used to improve animal welfare in UK farms. However, the analysis showed that the segment of consumers who are more likely to buy animal-friendly fresh pork are characterized by a high level of knowledge and awareness of the welfare of farm animals. Therefore, increasing consumers' knowledge about animal welfare and other desirable food attributes (e.g. organic, locality, health) could help improve consumers' interest in ethical food products and reduce their confusion when evaluating these attributes. This in turn is expected to increase consumers' demand for animal food products that carry individual, and bundles of, desirable food attributes.

3.3.3   Empirical study 2 – using market data to investigate demand for and substitutability/complementarity of pork products

In this study, market data on food and drink purchases for consumption at home were used to investigate: (i) the demand for animal-friendly and organic pork in Scotland; (ii) whether consumers in Scotland consider animal-friendly pork and organic pork as complementary products or substitutes; (iii) the relationship between pork products with different animal welfare labels (i.e. 'Freedom Food' pork versus 'Specially Selected Pork'); and (iv) whether the demand for animal-friendly and organic pork varies with the level of deprivation of the area where the household is located.

The empirical work used the Kantar Worldpanel data set, which includes weekly records of all foods and beverages that were taken home from supermarkets and similar stores by 3694 households during the period 2006–2011. The recruited households are representative of the Scottish population; however, not all of them were observed every year as the data set is a rotating panel (Hsiao, 2014). For each product, the data set contains rich information on a number of attributes (e.g. brand, organic, animal-friendly, origin). For each product, the data set also contains information on the price paid, the quantity purchased by the household, neighbourhood information and socio-demographic characteristics for all the households. The socio-economic information was expanded with information from the Scottish Neighbourhood Statistics, which among other variables, includes the Scottish Index of Multiple Deprivation (SIMD), the local authority and health board for the household.

To obtain the price and income elasticities, the linear version of the Almost Ideal Demand System (AIDS) was estimated. The model was estimated in three stages. In the first stage, aggregated household expenditure was first allocated among six groups of food products (i.e. beef, lamb, pork, chicken, fish and other foods). In the second stage, the expenditure on pork was allocated among four groups of pork products (i.e. fresh pork, bacon, ham and sausages). In the third stage, the expenditure on each group of pork products was allocated among four pork products (i.e. regular, organic, Freedom Food, Specially Selected Pork).

To compute the unconditional elasticities, the formulas reported in Carpentier and Guyomard (2001) were used. The model was estimated for the whole sample, and for two subsamples (SIMD 1 and SIMD 2). SIMD 1 includes all the households living in the more deprived areas and SIMD 2 includes all the households living in the less deprived areas. The analysis resulted in a large amount of results that cannot be included in this chapter but can be obtained from the authors upon request.

The estimation of the model was preceded by a descriptive analysis. The results from this analysis showed that, in Scotland, pork products are the most consumed meat, followed by chicken, beef, fish and lamb (11.18 kg, 9.15 kg, 5.76 kg, 3.30 kg and 0.82 kg per year per capita, respectively). Regarding the demand for animal-friendly pork, we found that only 5.6% of the total pork consumed in Scotland is labelled as animal-friendly (i.e. 'Freedom Food' or 'Specially Selected Pork'). Most of the consumed pork in Scotland is standard pork (93%) and only 0.02% of the total consumed pork is labelled as organic. It was also found that consumers in Scotland consume significantly more fresh pork, sausages, ham and bacon; however, the consumption of pork is 68% lower in more deprived areas than in less deprived areas. This gap increases to 77% in the case of animal-friendly pork.

The computed own-price elasticity is displayed in Table 3.2. All the own-price elasticities were negative and statistically different from zero, indicating that an increase in the price of any of the food products considered in the

Table 3.2. Unconditional own-price and expenditure elasticities.

Product	Own-price elasticities	Expenditure elasticities
Beef	−0.87528*	0.9020*
Lamb	−0.77433*	0.8163*
Chicken	−0.98297*	0.3272*
Fish	−1.20019*	0.2640*
Other foods	−0.99489*	−0.0173*
Fresh pork – regular	−1.29230*	0.1004*
Fresh pork – organic	−1.04022*	0.4911*
Fresh pork – FF	−1.90551*	0.1322*
Fresh pork – SSP	−0.34473*	−0.0988*
Bacon – regular	−1.30642*	1.2949*
Bacon cuts – organic	−0.89874*	0.3369*
Bacon – FF	−1.09117*	0.0992*
Bacon – SSP	−3.02253*	0.1754*
Ham – regular	−1.29171*	−0.6309*
Ham – organic	−0.95662*	0.3068*
Ham – FF	−1.07202*	0.2497*
Ham – SSP	−1.20952*	0.2839*
Sausages – regular	−1.11020*	0.3055*
Sausages – organic	−0.97783*	0.5282*
Sausages – FF	−0.94225*	0.9469*
Sausages – SSP	−1.00448*	1.0146*

Source: Author's elaboration based on Kantar Worldpanel data. FF stands for Freedom Foods and SSP stands for Specially Selected Pork.
Note: (*) Statistically significant at 5%

analysis will result in a decrease in its demand. The results also show that the demand for animal-friendly pork (i.e. 'Freedom Foods' and 'Specially Selected Pork') is somewhat elastic. This implies that an increase in its price by 1% will lead to a decrease in its demand by more than 1% (e.g. 1.9% for Freedom Food fresh pork, 1.09% for Freedom Food bacon, 3.02% for Specially Selected bacon, etc.) and vice versa. The results also show that the demand for animal-friendly pork is more sensitive to price variation than organic pork.

Interestingly, we found that the demand for animal-friendly pork products is more elastic in the more deprived areas. This implies that an increase in the prices of pork products certified as animal-friendly, for example as a result of new and more restrictive animal-welfare legislation, is likely to increase the disparity between the less and the more deprived areas in terms of pork consumption. This disparity could be further increased if new legislation on animal welfare results in an increase in the prices of the substitutes of animal-friendly pork (e.g. conventional pork) or banning its production altogether. It is possible that in the absence of affordable pork, more deprived consumers are likely to decrease their consumption of pork, which, in turn, may increase their food insecurity and widen the gap between them and rich consumers.

The unconditional cross-price elasticities are reported in Table 3.3. Overall, the results show that animal-friendly pork is competing with organic pork and, to a lesser extent, with conventional pork. This suggests that if the price of animal-friendly pork is increased, consumers are likely to substitute it with organic pork or conventional pork. Therefore, consumers' willingness to substitute animal-friendly pork with organic pork should be taken into account when considering increasing the price of animal-friendly pork due to, for example, a rise in the cost of production following a decision to comply with new animal welfare regulation. Furthermore, substituting animal-friendly pork with organic pork could also happen if consumers realize that organic pork is 'also' animal-friendly, especially when compared with conventional pork (Akaichi *et al.*, 2019).

We also found that 'Freedom Food' pork and 'Specially Selected' pork were substitutes, implying that an increase in the price of one of the two will lead to a decrease in its demand and an increase in demand for the other product. Another factor that may further increase the competition between these two pork products in Scotland is that the pork labelled as 'Specially Selected' is produced in Scotland. Since there is strong evidence in the literature that UK consumers are willing to pay a significant price premium for local food (e.g. Lobb *et al.*, 2006; Weatherell *et al.*, 2003; Akaichi *et al.*, 2019), labelling the Specially Selected Pork as 'local' could further increase its competitive power against 'Freedom Food' or organic pork.

The unconditional expenditure elasticities presented in Table 3.2 measure the sensitivity of the demand for the food products considered in the analysis to a variation in household income. The results show that consumers perceive animal-friendly pork as a 'normal' good. This indicates that an increase in household income by 1% will result in an increase in the demand for animal-friendly pork by less than 1%. Therefore the low expenditure elasticities for animal-friendly pork indicate that increasing households' income (e.g. subsidizing animal-friendly pork) may not be an effective strategy to boost the demand for this type of ethical pork.

3.4   Final Remarks

Farm animal welfare is a topic of increasing importance to producers, processers, retailers and consumers alike, so the debate on how to improve animal welfare in contemporary farms without putting their carers at financial risk is likely to be with us for some time. Economists can have an important role to play in shaping this debate.

The bulk of economic research on animal welfare in the last two decades has focused on investigating consumer demand for improved farm animal welfare. In this chapter, we presented and discussed examples of how economics can help in investigating the demand for animal welfare and how it is affected by drivers of food consumption such as price, income and non-price

Table 3.3. Unconditional cross-price elasticities.

	Fresh pork – regular	Fresh pork – organic	Fresh pork – FF	Fresh pork – SSP	Bacon – regular	Bacon – organic	Bacon – FF	Bacon – SSP
Fresh pork – regular	-0.3447*	-0.0542*	0.0000	0.0000	-0.0008*	-0.0203*	0.0000	0.0000
Fresh pork – organic	-0.0034*	-1.3059*	0.0005*	0.0001*	-0.0020*	-0.0485*	0.0000	0.0000
Fresh pork – FF	-0.0009	0.3342*	-0.8987*	0.1313*	-0.0032*	0.0113*	0.0000	0.0000
Fresh pork – SSP	0.0007	0.0537*	0.1063*	-1.0912*	-0.0026*	0.0095*	0.0000	0.0000
Bacon – regular	-0.0090*	-0.7699*	0.0175*	0.3232*	0.0002	-0.0012*	0.0000	0.0000
Bacon cuts – organic	0.0004	0.8002*	0.0003	-0.0003	-0.0010*	0.0037*	0.0000	0.0000
Bacon – FF	0.0001	0.2360*	0.0001	-0.0001	-0.0054*	0.5220*	0.0001	0.0002
Bacon – SSP	0.0002	0.4159*	0.0002	-0.0002	0.0795*	0.1408*	0.0000	0.0001
Ham – regular	-0.0006*	-1.5031*	-0.0006*	0.0006	1.1932*	-0.1045	0.0000	0.0000
Ham – organic	0.0235*	0.1222*	0.0001	0.0000	-3.0227*	1.3758*	0.0003*	0.0006*
Ham – FF	0.0191*	0.0992*	0.0000	0.0000	0.0143*	-1.2894*	0.0002*	0.0000
Ham – SSP	0.0218*	0.1131*	0.0000	0.0000	0.0042*	0.2651*	-0.9566*	0.3158*
Sausages – regular	0.0234*	0.1217*	0.0001	0.0000	0.0074*	-0.0648	0.3017*	-1.0720*
Sausages – organic	0.0016*	0.1224*	0.0000	0.0000	-0.0268*	0.5792*	-0.0841*	3.1362*
Sausages – FF	-0.0019*	-0.1472*	-0.0001	0.0001	0.0021*	-0.0507*	0.0000	0.0000
Sausages – SSP	0.0000	0.0021*	0.0000	0.0000	0.0017*	-0.0413*	0.0000	0.0000
Fresh pork – regular	0.0000	0.2673*	0.0004	0.0024*	0.0227*	0.0606*	-0.0264*	0.0934*
Fresh pork – organic	0.0000	0.0892*	0.0002	0.0008*	0.0078*	0.2965*	-0.1290*	0.4564*
Fresh pork – FF	0.0000	0.0242*	0.0000	0.0002	0.0021*	0.0798*	-0.0347*	0.1233*
Fresh pork- SSP	0.0000	-0.0178*	0.0000	-0.0002	-0.0015*	-0.0591*	0.0260*	-0.0915*
Bacon – regular	-0.0001	0.2353*	0.0004	0.0021*	0.0205*	0.7820*	-0.3405*	1.2006*
Bacon cuts – organic	0.0000	-0.0577*	-0.0001	-0.0005	-0.0048*	0.2032*	-0.0884*	0.3131*
Bacon – FF	-0.0046*	-0.0170*	0.0000	-0.0002	-0.0014*	0.0598*	-0.0261*	0.0922*
Bacon – SSP	0.1650*	-0.0300*	0.0000	-0.0003	-0.0025*	0.1057*	-0.0460*	0.1631*
Ham – regular	-1.2095*	0.1081*	0.0002	0.0010	0.0089*	-0.3811*	0.1653*	-0.5866*

Continued

Table 3.3. Continued

	Fresh pork – regular	Fresh pork – organic	Fresh pork – FF	Fresh pork – SSP	Bacon – regular	Bacon – organic	Bacon – FF	Bacon – SSP
Ham – organic	0.0863*	−1.1097*	−0.0001	−0.0010*	−0.0092*	0.1848*	−0.0804*	0.2851*
Ham – FF	0.0000	−0.0422*	−0.9778*	0.1003*	0.0089*	0.1504*	−0.0654*	0.2320*
Ham – SSP	0.0000	−0.0974*	0.0209*	−0.9423*	−0.0173*	0.1709*	−0.0744*	0.2638*
Sausages – regular	0.0000	−0.1092*	0.0002	−0.0020*	−1.0042*	0.1840*	−0.0801*	0.2839*
Sausages – organic	0.0000	0.0554*	0.0001	0.0005	0.0046*	−0.9774*	0.2930*	0.1786*
Sausages – FF	0.0000	−0.0667*	−0.0001	−0.0006*	−0.0056*	0.8104*	−1.1966*	0.7678*
Sausages – SSP	0.0000	0.0009*	0.0000	0.0000	0.0001*	0.0019*	0.0030*	−0.0157*

Source: Author's elaboration based on Kantar Worldpanel data. FF stands for Freedom Foods and SSP stands for Specially Selected Pork.
Note: (*) Statistically significant at 5%

food attributes (e.g. origin, organic). However, animal welfare is a complex topic and improving the wellbeing of farm animals is not only determined by consumer awareness and behaviour but also by other stakeholders' decisions (e.g. farmers, retailers, policy makers).

For instance, farmers' management decisions and day-to-day husbandry determine, to a large extent, their animals' welfare. Therefore, understanding farmer decision making and its determining factors is crucial to develop strategies and techniques to improve animal welfare within the farm. Economic research can contribute to gaining better understanding of farmers' behaviour and its drivers, as well as identifying incentives to motivate farmers to safeguard and improve animal welfare.

Animal welfare is a good example of a public good. This implies that the knowledge of improved animal welfare can be enjoyed by all those in the society and not just those who care about animal welfare and who are willing to pay a price premium for it. This is one of the reasons that make producers and marketers of food products reluctant to provide products with a high level of public good such as animal welfare. However, because improving animal welfare can benefit a large part of society, the government usually intervenes to protect animal welfare through, for example, legislation, subsidies to farmers committed to improving animal welfare, provision of information for consumers on farm animal welfare and helping farmers to improve their farming, business and marketing skills. Economics can be used to analyse the effectiveness of these policy decisions and help policy makers identify and target different segments of stakeholders with strategies tailored to their needs.

Note

[1] See Deaton and Muellbauer (1980) for more details about the demand model's assumptions and how the elasticities are derived. The unconditional price and income elasticities can be estimated using the approach described in Carpentier and Guyomard (2001).

References

Akaichi, F., Glenk, K. and Revoredo-Giha, C. (2019) Could animal welfare claims and nutritional information boost the demand for organic meat? Evidence from non-hypothetical experimental auctions. *Journal of Cleaner Production* 207, 961–970. DOI: 10.1016/j.jclepro.2018.10.064.

Baltzer, K. (2004) Consumers' willingness to pay for food quality – the case of eggs. *Acta Agriculturae Scandinavica, Section C — Food Economics* 1(2), 78–90. DOI: 10.1080/16507540410024506.

Bennett, R.M. (1997) Farm animal welfare and food policy. *Food Policy* 22(4), 281–288. DOI: 10.1016/S0306-9192(97)00019-5.

Blount, W.P. (1968) Factory farming – animal machines? In: Blount, W.P. (ed.) *Intensive Livestock Farming*. Heinemann Medical Books, London, pp. 522–556.

Carlsson, F., Frykblom, P. and Lagerkvist, C.J. (2007a) Consumer willingness to pay for farm animal welfare: mobile abattoirs versus transportation to slaughter. *European Review of Agricultural Economics* 34(3), 321–344. DOI: 10.1093/erae/jbm025.

Carlsson, F., Frykblom, P. and Lagerkvist, C.J. (2007b) Farm animal welfare—testing for market failure. *Journal of Agricultural and Applied Economics* 39(1), 61–73. DOI: 10.1017/S1074070800022756.

Carpentier, A. and Guyomard, H. (2001) Unconditional elasticities in two-stage demand systems: an approximate solution. *American Journal of Agricultural Economics* 83(1), 222–229. DOI: 10.1111/0002-9092.00149.

Chang, J.B., Lusk, J.L. and Norwood, F.B. (2010) The price of happy hens: a hedonic analysis of retail egg prices. *Journal of Agricultural and Resource Economics* 35(3), 406–423.

Cronin, G.M., Rault, J.L. and Glatz, P.C. (2014) Lessons learned from past experience with intensive livestock management systems. *Revue Scientifique et Technique de l'OIE* 33(1), 139–151.

Deaton, A. and Muellbauer, J. (1980) An almost ideal demand system. *The American Economic Review* 70(3), 312–326.

European Commission (2005) Attitudes of consumers towards the welfare of farmed animals. Available at: https://ec.europa.eu/food/sites/food/files/animals/docs/aw_arch_euro_barometer25_en.pdf (accessed 17 June 2020).

European Commission (2007) Attitudes of EU citizens towards animal welfare. Available at: http://www.vuzv.sk/DB-Welfare/vseob/sp_barometer_aw_en.pdf (accessed 17 June 2020).

Gracia, A., Loureiro, M.L. and Nayga Jr., R.M. (2011) Valuing an EU animal welfare label using experimental auctions. *Agricultural Economics* 42(6), 669–677. DOI: 10.1111/j.1574-0862.2011.00543.x.

Gracia, A., Barreiro-Hurlé, J. and Galán, B.L. (2014) Are local and organic claims complements or substitutes? A consumer preferences study for eggs. *Journal of Agricultural Economics* 65(1), 49–67. DOI: 10.1111/1477-9552.12036.

Grunert, K.G. (2006) Future trends and consumer lifestyles with regard to meat consumption. *Meat Science* 74(1), 149–160. DOI: 10.1016/j.meatsci.2006.04.016.

Harper, G. and Henson, S. (2001) *Consumer concerns about animal welfare and the impact on food choice*. Final project report, European Commission FAIR programme CT98-3678. Centre for Food Economics Research, Department of Agricultural & Food Economics, Reading University, UK.

Hensher, D.A., Rose, J.M. and Greene, W.H. (2015) *Applied Choice Analysis: A Primer*. Cambridge University Press, Cambridge, UK.

Hill, H. and Lynchehaun, F. (2002) Organic milk: attitudes and consumption patterns. *British Food Journal* 104(7), 526–542. DOI: 10.1108/00070700210434570.

Honkanen, P. and Ottar Olsen, S. (2009) Environmental and animal welfare issues in food choice: the case of farmed fish. *British Food Journal* 111(3), 293–309.

Hsiao, C. (2014) *Analysis of Panel Data*. Vol. 54. Cambridge University Press, Cambridge, UK.

Hutchins, R.K. and Greenhalgh, L.A. (1997) Organic confusion: sustaining competitive advantage. *British Food Journal* 99(9), 336–338.

Kanis, E., Groen, A.F. and De Greef, K.H. (2003) Societal concerns about pork and pork production and their relationships to the production system. *Journal of Agricultural and Environmental Ethics* 16(2), 137–162. DOI: 10.1023/A:1022985913847.

Kehlbacher, A., Bennett, R. and Balcombe, K. (2012) Measuring the consumer benefits of improving farm animal welfare to inform welfare labelling. *Food Policy* 37(6), 627–633. DOI: 10.1016/j.foodpol.2012.07.002.

Lagerkvist, C.J., Carlsson, F. and Viske, D. (2006) Swedish consumer preferences for animal welfare and biotech: a choice experiment. *AgBioForum* 9(1), 51–58.

Lagerkvist, C.J. and Hess, S. (2011) A meta-analysis of consumer willingness to pay for farm animal welfare. *European Review of Agricultural Economics* 38(1), 55–78. DOI: 10.1093/erae/jbq043.

Lancaster, K.J. (1966) A new approach to consumer theory. *Journal of Political Economy* 74(2), 132–157.

Liljenstolpe, C. (2008) Evaluating animal welfare with choice experiments: an application to Swedish pig production. *Agribusiness* 24(1), 67–84. DOI: 10.1002/agr.20147.

Lobb, A.E., Arnoult, M.H., Chambers, S. and Tiffin, R. (2006) *Willingness to pay for, and consumers' attitudes to, local, National and imported foods: a UK survey.* University of Reading, Berkshire.

Lusk, J.L. (2010) The effect of proposition 2 on the demand for eggs in California. *Journal of Agricultural & Food Industrial Organization* 8(1), 1–20. DOI: 10.2202/1542-0485.1296.

Lusk, J.L. and Norwood, F.B. (2008) A survey to determine public opinion about the ethics and governance of farm animal welfare. *Journal of the American Veterinary Medical Association* 233(7), 1121–1126. DOI: 10.2460/javma.233.7.1121.

Lo, M. and Matthews, D. (2002) Results of routine testing of organic food for agro-chemical residues. In: *Proceedings of the UK Organic Research 2002 Conference.* Organic Centre Wales, Institute of Rural Studies, University of Wales, Aberystwyth, UK, pp. 61–64.

Lusk, J.L., Norwood, F.B. and Pruitt, J.R. (2006) Consumer demand for a ban on antibiotic drug use in pork production. *American Journal of Agricultural Economics* 88(4), 1015–1033. DOI: 10.1111/j.1467-8276.2006.00913.x.

Lusk, J.L., Norwood, F.B. and Prickett, R.W. (2007) *Consumer Preferences for Farm Animal Welfare: Results of a Nationwide Telephone Survey.* Oklahoma State University, Department of Agricultural Economics, Stillwater, Oklahoma.

Mayfield, L.E., Bennett, R.M., Tranter, R.B. and Wooldridge, M.J. (2007) Consumption of welfare-friendly food products in Great Britain, Italy and Sweden, and how it may be influenced by consumer attitudes to, and behaviour towards, animal welfare attributes. *International Journal of Sociology of Agriculture and Food* 15(3), 59–73.

McFadden, D. (1974) Conditional Logit analysis of qualitative choice behavior. In: Zarembka, P. (ed.) *Frontiers in Econometrics.* Academic Press, New York, pp. 105–142.

Nocella, G., Hubbard, L. and Scarpa, R. (2010) Farm animal welfare, consumer willingness to pay, and trust: results of a cross-national survey. *Applied Economic Perspectives and Policy* 32(2), 275–297. DOI: 10.1093/aepp/ppp009.

Norwood, F.B. and Lusk, J.L. (2011a) Compassion, by the pound: the economics of farm animal welfare. *OUP Catalogue.*

Norwood, F.B. and Lusk, J.L. (2011b) A calibrated auction-conjoint valuation method: valuing pork and eggs produced under differing animal welfare conditions. *Journal of Environmental Economics and Management* 62(1), 80–94. DOI: 10.1016/j.jeem.2011.04.001.

Raun Mørkbak, M., Christensen, T. and Gyrd-Hansen, D. (2010) Consumer preferences for safety characteristics in pork. *British Food Journal* 112(7), 775–791. DOI: 10.1108/00070701011058299.

Schroder, M.J.A. and McEachern, M.G. (2004) Consumer value conflicts surrounding ethical food purchase decisions: a focus on animal welfare. *International Journal of Consumer Studies* 28(2), 168–177. DOI: 10.1111/j.1470-6431.2003.00357.x.

Street, D.J. and Burgess, L. (2007) *The Construction of Optimal Stated Choice Experiments: Theory and Methods.* Vol. 647. Wiley, Chichester, UK.

Toma, L., Kupiec-Teahan, B., Stott, A.W., Revoredo-Giha, C., Darnhofer, I. *et al.* (2010) Animal Welfare, Information and Consumer Behaviour. *9th European IFSA Symposium*, pp. 4–7.

Toma, L., McVittie, A., Hubbard, C. and Stott, A.W. (2011) A structural equation model of the factors influencing British consumers' behaviour toward animal welfare. *Journal of Food Products Marketing* 17(2–3), 261–278. DOI: 10.1080/10454446.2011.548748.

Tonsor, G.T. and Olynk, N.J. (2011) Impacts of animal well-being and welfare media on meat demand. *Journal of Agricultural Economics* 62(1), 59–72. DOI: 10.1111/j.1477-9552.2010.00266.x.

Tonsor, G.T., Olynk, N. and Wolf, C. (2009) Consumer preferences for animal welfare attributes: the case of gestation crates. *Journal of Agricultural and Applied Economics* 41(3), 713–730. DOI: 10.1017/S1074070800003175.

Train, K. (1998) Recreation demand models with taste differences over people. *Land Economics* 74(2), 230–239. DOI: 10.2307/3147053.

Train, K. (2003) *Discrete Choice Methods with Simulation.* Cambridge University Press, Cambridge, UK.

Train, K. and Weeks, M. (2005) Discrete choice models in preference space and willingness-to-pay space. In: Scarpa, R. and Alberini, A. (eds) *Applications of Simulation Methods in Environmental and Resource Economics.* Springer, Dordrecht, The Netherlands, pp. 1–16.

Vanhonacker, F., Van Poucke, E., Tuyttens, F. and Verbeke, W. (2010) Citizens' views on farm animal welfare and related information provision: Exploratory insights from Flanders, Belgium. *Journal of Agricultural and Environmental Ethics* 23(6), 551–569. DOI: 10.1007/s10806-010-9235-9.

Verbeke, W. (2009) Stakeholder, citizen and consumer interests in farm animal welfare. *Animal Welfare* 18(4), 325–333.

Verbeke, W., Van Oeckel, M.J., Warnants, N., Viaene, J. and Boucqué, C.V. (1999) Consumer perception, facts and possibilities to improve acceptability of health and sensory characteristics of pork. *Meat Science* 53(2), 77–99. DOI: 10.1016/S0309-1740(99)00036-4.

Weatherell, C., Tregear, A. and Allinson, J. (2003) In search of the concerned consumer: UK public perceptions of food, farming and buying local. *Journal of Rural Studies* 19(2), 233–244. DOI: 10.1016/S0743-0167(02)00083-9.

4 People's Preferences in Relation to Animal Welfare

RICHARD BENNETT*

School of Agriculture, Policy and Development, University of Reading

Summary

People may have preferences in relation to products and services that they perceive to have animal welfare attributes. People may also have preferences in relation to different welfare states, standards and treatment of animals. Human empathy with animals is a potential source of utility (e.g. the pleasure from seeing animals playing) and disutility (e.g. the distress caused if we perceive an animal is suffering). We may derive satisfaction from moral preferences for how animals are treated regardless of this empathy. There are many animal-derived products for consumers to purchase in markets but little information on the welfare of the animals that produced them in order for consumers to make choices consistent with their animal welfare preferences. People's preferences can be ascertained by observations of their behaviour and the choices they make, and by asking people what their preferences are. Two research case studies are presented. There is a need for welfare advocacy on behalf of animals. This stewardship role must be informed by scientific evidence on animal sentience, preferences and welfare status. Government or an alternative body acting on behalf of society must take up the role to represent societal animal welfare preferences and act as custodians to protect the welfare of animals.

4.1 People's Preferences and Utility

The concept of preference is an important one in consumer theory and welfare economics. In economic theory, individuals are assumed to have the objective of maximizing their utility (assumed synonymous with their happiness and welfare) and make decisions based on expected utility maximization.

*r.m.bennett@reading.ac.uk

It is further assumed that people maximize their expected utility by acting on their preferences (preference utilitarianism – see Bekoff (2009) for a succinct description and discussion in relation to animal welfare) and choosing particular goods (and services) or bundles of goods in preference to others which have a lower expected utility for the decision maker. Choice is therefore essential to be able to express preference, and choice behaviour is interpreted as an indication of people's preferences. Expression of preferences, and choice behaviour, is greatly facilitated by, and observed in, markets. This requires availability and knowledge of alternatives from which to choose. For example, in food retail outlets there is a vast array of food products with different attributes for consumers to choose from with various information on ingredients, nutritional quality and other aspects.

So how does the concept of preference relate to animal welfare?[1] People may have preferences in relation to various products and services that they perceive to have animal welfare attributes or that are associated with a particular animal welfare provenance. People may also have preferences in relation to different welfare states, standards and treatment of animals. In the context of animal welfare, the problem is that, both as consumers and as citizens, we have limited opportunities to express our preferences in relation to animal welfare (e.g. in relation to markets and democratic processes). These preferences may relate to animals used in food and fibre production, production of health and cosmetic products, companion animals, working animals used, for example, for traction and transport, animals used in research, in zoos, for entertainment and sport, and wild animals which may be partially managed (e.g. nature reserves) or not managed by humans (but may be affected by human activities, for example in relation to habitat loss).

The basis of these preferences is expected utility and it is assumed that we act to maximize this utility. So what are the sources of utility or satisfaction that we can derive in relation to animal welfare? The first category of utility source in relation to animal welfare is that associated with an expectation that products with a high animal welfare provenance are in some sense better than those associated with lower welfare of animals. Thus the perceived enhanced products (due to better taste, quality, etc.) provide greater expected utility of consumption (see Bennett *et al.*, 2012). The second source relates to our empathy with animals and belief in/understanding of their sentience. In human evolutionary terms, this empathy possibly relates both to the evolution of empathy for other humans (especially babies) leading to greater care and survival of these humans, which has then also transferred to empathy for animals (people often have greatest empathy for young animals), and evolution of empathy with animals *per se*, possibly because this type of empathy may lead to greater understanding of animals' feelings and motivations, which may help humans to better predict animal actions and behaviour and so be better hunters of those animals. Human empathy with the feelings of animals is a potential source of utility (the pleasure we may derive from seeing animals apparently joyfully playing or basking in the sun) and disutility (the distress we may feel if we witness an animal suffering or perceive an animal to be suffering, for example from an injury or from disease). Arguably, if we did not have empathy with animal feelings then the only source of utility and preference in relation to animal welfare would be the expected tangible benefits associated with animal-derived products and services (although we could conceivably still have moral preferences for how we treat animals regardless of our empathy with them (see below)). The second source of utility associated with animal welfare, empathy with animal feelings, is probably the most important and the most powerful in influencing our preferences – both as consumers and citizens – given the nature of its embeddedness in the human condition.

There is a third possible category or source of utility and preference in relation to animal welfare, which is associated with people acting as moral agents (long recognized by a number of moral philosophers and economists including Smith, 1790; Bentham, 1789; and Marshall, 1947). There is a certain satisfaction (utility) that can be derived from feeling that you are 'doing the right thing', or acting in the right way'. This is linked to enhanced feelings of positive self-esteem, which is considered by many to be a major source of positive feeling for people and driver of human behaviour (see Zhang, 2009, on money, self-esteem and utility). In addition, there is potential satisfaction to be gained, a social satisfaction, from displaying moral values and behaviour which is also linked to the satisfaction derived from having affinity with others. Bentham (1789) notes the utility derived from sympathy with others, while Mill, 1863 expands the concept of utility and utilitarianism to include the 'pleasures' of higher moral values, including selflessness, duty and other moral behaviours. Given the complexity of the sources of potential utility/disutility in relation to animal welfare, it is no wonder that our preferences for animal welfare are often difficult to understand and gauge.

4.2   Difficulty for Consumers of Expressing Animal Welfare Preferences in Markets

Although there are many animal-derived products available for consumers to purchase in markets, there is often little or no information relating to the welfare of the animals that produced those products, thus enabling consumers to make purchasing choices consistent with their animal welfare preferences. Products which have some reference regarding animal welfare include free-range and organic eggs and meat, and, more specifically in relation to animal welfare, products which are labelled as RSPCA Assured, Animal Welfare Approved (Sustainable Food Trust) and Global Animal Partnership certified. However, these designations provide varying and different standards and assurances in relation to the welfare of the animals used to produce those products. In the case of a free-range designation, this provides some (limited) information relating to the system of egg production but nothing directly about the welfare of the hens involved. Free-range production systems can also vary considerably. It may be considered that free-range systems have a greater potential for high animal welfare compared to, for example, caged systems, but there is no assurance of this for the consumer. Consumers may also purchase free-range or other products which have some perceived animal welfare credentials because they believe them to be of higher quality, better for the environment, safer for the consumer and taste better, as well as causing less or no suffering to the animals that produced them, compared with intensively housed animals. Thus, consumers may purchase goods with perceived animal welfare attributes for reasons associated with the wellbeing/utility of the animals concerned. For example, in a focus group of meat purchasers undertaken by the author, one of the participants said that they preferred to eat meat with a high welfare provenance because they were concerned that animals that were stressed and had poor welfare produced hormones in their bodies which then could be transferred to their meat (see FAO, 2001 for evidence that animals stressed at slaughter have hormonal abnormalities which affect the quality of their meat).

Walking around the food store (or searching food retailers' offerings online), it is very difficult for consumers to find information or even clues regarding the animal welfare attributes or provenance of products. It is even more difficult where food service is concerned, such as food in restaurants, takeaway food, employer cafeterias, etc. Thus the consumer is greatly constrained and inhibited in terms of her/his ability to express any preferences for animal-based products that come from animals experiencing high levels of wellbeing throughout their lives. Consumers are left having to largely trust food retailers and their suppliers that animals have been treated humanely. This lack of information about animal welfare is a major problem preventing consumers from making informed choices regarding the products they purchase and preventing them from expressing their preferences in relation to animal welfare.

4.3   Preferences as Citizens

Consumers are also citizens (together with non-consumers of some or all animal-derived products). As citizens we may have different sets of preferences compared to our preferences as consumers. For example, as a citizen I may feel that all animals should have good welfare with minimal suffering throughout their lives and may decide to vote for legislation which encapsulates this belief if given the opportunity. However, as a consumer, I may, on occasion at least, purchase products that are not associated with high levels of animal welfare such as eggs produced from hens kept in cages. Such purchases by those professing to care about animals are often cited by producer organizations and food industry representatives as evidence that consumers do not really care about animal welfare; they just say that they do. Of course, the reality is that, as both citizens and consumers, we have many different preferences, competing demands and conflicting beliefs. Thus, I may have chosen to purchase the caged eggs because of poor or even misleading information – I thought when they said 'farm assured' that they were free-range, or perhaps my budget that week was highly constrained and so

I purchased the cheapest products available. Consumers and others may not have stable preferences over time or may have seemingly inconsistent preferences because of lack of information and other competing demands/preferences. Indeed, our beliefs as citizens and our actions as consumers may result in significant cognitive dissonance (a feeling of unease associated with inconsistent beliefs, attitudes or behaviour) when they are not in accord. For some, this cognitive dissonance may become sufficiently strong to change their purchasing behaviour (e.g. buy free-range products, become vegetarian or vegan), whereby the utility gained from consumption is outweighed by the disutility of the feelings associated with cognitive dissonance (Bennett, 1995). Arguably, as consumers, we are seeking to maximize our expected utility from the purchase and consumption of the products we buy, whereas as citizens we are more seeking to do what is right – i.e. we see ourselves in a more moral role as citizens – or perhaps in an economic utilitarian context we are seeking to maximize our expected utility over a greater range of considerations, which includes both utility from consumption and utility derived from satisfying 'higher needs' such as acting according to moral values or codes of conduct. This means that as citizens we may vote for something that may reduce our own and others' consumer-related utility with the assumption that this reduction in utility from consumption is outweighed by the satisfaction/utility of knowing that we have done 'the right thing'. Greater information regarding animal welfare may increase cognitive dissonance for consumers and, in the short run, reduce their utility, at least until they adjust their behaviour to take into account the new information that they have. However, this reduction in utility should not be used as a reason to deprive consumers of the information they need.

4.4 Animal Preferences and Utility

There is an implicit assumption in economics that it is human preferences and human utility that matter and not animal preferences or animals' utility and that the latter only matter as a subset of people's utility functions (McInerney, 1994). However, many (including Bentham, 1789; Singer, 1975) would consider that the welfare of animals is important in its own right and that the utility of all sentient beings should be taken into account. Although this view may give rise to some 'moral satisfaction' for people who support it, the principle concerns animal and not human utility. There is an increasing scientific literature in relation to animal preferences and better understanding of animal utility related to significant scientific advances in understanding animal sentience. The science of assessing and measuring animal preferences has developed over the last 20 or so years in particular. It has evolved from merely observing the apparent choices made by animals in relation to their environment to a range of experimental methods to measure animal preferences in different contexts. As with human preferences, animal preferences have been found to be complex and so increasingly complex research methods are being developed to better understand those preferences, what lies behind them (e.g. the strength of motivation) and the importance of context (e.g. in terms of the animal's environment, previous experiences, condition/ state, cognitive capacity, etc.). Animal behaviour, preferences and animal demand have been variously considered using economic principles by Hursh (1980) and Dawkins (1990) among others. The reader is referred to Fraser and Nicol (2018) for a useful introduction to animal preference measurement. Dawkins (1990) applied the economic concept of demand elasticity for estimating the motivational strength associated with animal preferences. The 'price' variable in such studies is usually in relation to the amount of time or effort expended to gain access to alternative environments, foodstuffs, etc.

Arguably, the more that we, as citizens, understand and appreciate the sentience of animals, their feelings, preferences and needs, the stronger our empathy with them and hence the stronger our own preferences may become to pursue higher levels of welfare for animals. Despite this understanding, we are often somewhat 'speciesist' (Regan, 1982) in this regard. For example, we are generally less concerned with the welfare of rats, which have relatively high sentience, but we treat them as vermin, traditionally dosing them with poisons such as warfarin (which likely leads to a relatively long and painful death), compared with other (perhaps less sentient) species which have charitable societies set up to protect their welfare. However, if sentience is the basis of the utilitarian ethic and utilitarianism is the bedrock of the discipline of economics, then, arguably, the principles of economic theory (and the utilitarian ethic) in relation to human welfare and maximization of utility should extend to all sources of utility and hence to all sentient beings. This argument supports the assertion by Singer (1975) of the need for 'equal consideration of interests' of humans and animals. Singer writes, 'If a being suffers, there can be no moral justification for refusing to take that suffering into consideration'. But the sentience and preferences of animals do not (directly) form part of our utilitarian metric. Hence, Bentham (1789) noted that 'animals, which, on account of their interests having been neglected ... stand degraded into the class of things ...'.

However, there are a number of theoretical, philosophical and practical problems associated with using measures of animal preferences to infer what the satisfaction of these apparent preferences mean for animal utility and the welfare of animals. First, animals have limited opportunity to express their preferences and human society has limited ability/opportunity to gauge them, certainly at the individual animal level. Instead, broad generalizations are made by those in human society about the type of environment, food, behaviours, etc. that particular species or animals in particular systems might prefer. These generalizations will often not be accurate for a specific individual or under particular circumstances or contexts. In addition, as for humans, animals' preferences may not always lead to good welfare outcomes, and for both animals and some humans (e.g. the very young, those with cognitive impairment etc.), decisions regarding what is in the best welfare interests of those groups and individuals are often taken by others in a guardianship role. There is also the issue as to how we weigh the preferences of one individual or group against those of others. In human society we rely on markets and democratic processes such as voting systems, representation, lobbying, right to protest – all of which are denied to animals. Animal representation and advocacy are undertaken in human society on behalf of animals by a number of animal interest groups such as the charitable organization People for Ethical Treatment of Animals (PETA). Many of these organizations express preferences purportedly on both animals' and citizens' behalf. Animal welfare interest groups such as the Royal Society for the Prevention of Cruelty to Animals (RSPCA) and Compassion in World Farming (CIWF) in the UK express preferences on behalf of citizens and consumers – the assumption being that people would choose to express these preferences in relation to animal welfare if they were able or mobilized to do so. Governments often act similarly on society's behalf to protect the welfare of animals via codes of practice, legislation and other means.

The reason for this advocacy is that animal preferences can only be incorporated into a social welfare function in a theoretical way (see Blackorby and Donaldson, 1992; Johansson-Stenman, 2018) as there is no practical means for doing so. As noted earlier, animals have no vehicle to express their preferences in a (human) societal context. The implications of this for animal welfare policy are considered towards the end of the chapter.

4.5   Understanding and Measuring (Human) Preferences

People's preferences can be ascertained/measured in two main ways. The first involves observations of people's behaviour and the choices they make. This is made easier by being able to observe and record people's behaviour in markets in terms of what they buy, how much they pay and the context of this behaviour (the choice environment, people's income and other characteristics). Of course, there are many choices that we make with underlying preferences that are not represented by market transactions and so we can only infer a limited set of preferences from market data. The second way of measuring preferences is to ask people what their preferences are. People often do this on a daily basis in conversation with each other either directly or indirectly – and such conversations can now be more easily mapped from comments and interactions on social media. Other sources of information regarding our preferences include how people vote (for example, the 2018 California Proposition 12 Farm Animal Confinement Initiative, which was passed with more than 61% of the vote, setting specific legal space requirements for confined animals raised for food), the communications sent to government and other officials (animal welfare issues have been mentioned in a relatively large share of letters of concern to UK MPs) and donations to animal welfare charities. Another chapter in this book considers consumer behaviour and market demand. Here we focus on the stated preference way of gauging preferences.

There is an extensive literature on stated preference techniques used to elicit people's preferences and the value they give to particular choices (e.g. Louviere *et al.*, 2000). There is a much less, but growing, literature in the area of animal welfare. Bennett *et al.* (2011) provide a review and critique of valuation studies applied to animal welfare. Lagerkvist and Hess (2011) identify 24 stated preference studies of people's willingness to pay (WTP) in relation to animal welfare, yielding 106 WTP estimates, in their meta-analysis. Approximately half of these are contingent valuation studies and half use a choice experiment approach (with one other using an experimental auction method). A common feature of these studies is that they elicit preferences and/or willingness to pay (WTP) from citizens regarding proposed specific changes in animal production practices or other measures to improve welfare, often using hypothetical scenarios. WTP values are then used as estimates of the likely magnitude of the benefits that citizens obtain from each of these changes in practices or other measures.

A broad indication of people's attitudes, and to some extent their likely preferences, is provided by surveys of opinions elicited from a sample of citizens. The European Commission's Eurobarometer survey reporting on attitudes to animal welfare across Europe is a useful indicator of attitudes over time. The 2015 survey covering 28 countries and involving 27,672 citizen respondents (European Commission, 2016a) reported that 94% of respondents said that protecting the welfare of farm animals is important; 89% believe there should be EU legislation that requires people to care for animals used for commercial purposes; 82% think farm animals should be better protected than they are now; 74% said that companion animals need greater protection; and 59% are willing to pay 5% more for animal-friendly products.

Two case studies are presented below as examples of how people's preferences in relation to animal welfare can be elicited.

Case Study 1, and studies like it, pose a number of potential problems. The first, which is common to all stated preference studies, is that there are numerous sources of potential bias of the WTP measure at all stages of the elicitation and estimation process. For example, Loomis (2014) reviews problems related to hypothetical biases and considers both *ex ante* and *ex post* strategies for overcoming them.

The second potential drawback of the approach used in Case Study 1 is that the estimation WTPs are only relevant to the precise context of the measures involved. This criticism applies more generally to all previous such WTP studies applied to measures intended to improve animal welfare. Thus, people's WTP from one intervention or policy is not transferable to other policies or changes in animal husbandry. Moreover, in studies to estimate people's WTP for the interventions, neither the interventions themselves nor people's WTP to support them are linked to explicit animal welfare changes. Given that there are, potentially, thousands of different interventions to improve the welfare of animals across different species, production systems, etc. this would necessitate thousands of such studies to estimate the WTP and value that citizens give to the benefits likely to accrue from those interventions. Moreover, the validity of each of these estimates may be in question because the different studies are likely to be undertaken in different ways and not validated through being repeated (since each study and each estimate is unique). This makes policy evaluation very costly, cumbersome

Case Study 1: People's willingness to pay for improved meat chicken welfare

In 2007, new EU rules were agreed (Council Directive 2007/43/EC) for protecting the welfare of broiler chickens (European Commission Council Directive 2007/43/EC, 2007). The broiler directive came into force in the UK on 30 June 2010. Expected benefits of the legislation included consumer perceptions of enhanced chicken meat quality and enhancement of consumers' and citizens' levels of satisfaction from knowing that the welfare of broiler chickens is better protected in the food production process (Defra, 2010).

A study was undertaken for the Department of Food and Rural Affairs (Defra), which tested the hypothesis that citizens have an additional WTP for the broiler directive and assessed the scale of such benefits by means of an *ex post* WTP survey of citizens.

In the UK, the broiler directive has been implemented such that producers have been permitted to choose from among two stocking density allowances:

1. Stocking up to 33 kg live weight/m² if specific standards are met for drinkers, feeding, litter, ventilation and heating, noise, light, inspections, cleaning, record keeping, training and surgical interventions;
2. Stocking beyond 33 kg/m² up to 38 kg/m² if a specific set of additional standards is met. These standards include notification and documentation requirements, plus further controls on environmental parameters in broiler housing.

In addition, the directive requires the collection of data from farms on cumulative daily mortality (CDM) and data from slaughterhouses for eight post-mortem measures of body condition (collectively known as the 'trigger conditions') to help identify poor welfare on farms. These data, which relate to each batch of birds per farm sent to slaughter, are used by the Food Standards Agency (FSA) and the Animal and Plant Health Agency (APHA) to identify farms that may require problem notification and/or on-site inspection. The post-mortem body condition measures for which data are collected are: (i) ascites/oedema; (ii) cellulitis and dermatitis; (iii) dead on arrival; (iv) emaciation; (v) joint lesions/arthritis; (vi) septicaemia/respiratory problems; (vii) total carcass rejections; and (viii) foot pad dermatitis score. Evaluation of these data involves two processes:

Process 1: An alert (to APHA) is triggered if the incidence of any of the individual post-mortem conditions is exceptionally high in any batch (defined as greater than six standard deviations above the mean).

Process 2: An alert is triggered if the CDM is unusually high (defined as greater than three standard deviations above the mean) and, additionally, the level of three or more of the post-mortem conditions is high (defined as above the mean).

When trigger thresholds are breached, the keeper of the animals and APHA are alerted by means of a 'trigger report'. APHA and FSA have inspection regimes and data handling systems to communicate information relating to poor welfare between the slaughterhouse and the producer. Investigative action will be taken by APHA veterinary officers, and this may include requesting a written action plan to remedy the problem from producers and/or a visit to the production site. APHA may, in addition, carry out a number of random welfare inspections.

A questionnaire was designed which contained an introduction explaining the nature of the survey and its purpose, followed by questions to respondents with regards to their:

- personal characteristics (sample stratification variables);
- current consumption of chicken meat;
- attitudes towards farmed animal welfare;

Continued

Continued.

> - WTP for farmed animal welfare improvements in general;
> - attitudes towards the broiler directive;
> - WTP for the directive and debriefing questions to explain respondents' WTP responses;
> - socio-demographic characteristics (non-stratification variables).
>
> Chicken consumers were asked how much extra they would be willing to pay for the directive in the form of a premium on the price of the conventionally reared chicken products that they purchased, while non-chicken eaters were asked their WTP as an additional sum on their income tax. The survey was carried out by means of a web-based questionnaire and a stratification procedure was applied during recruitment of survey participants to ensure that the sample was broadly representative of populations in England and Wales.
>
> Respondent WTP was elicited using a contingent valuation (CV) payment card with a discrete dichotomous choice format (Mitchell and Carson, 1989) with multiple increasing values akin to a 'bidding game' (e.g. Heinzen and Bridges, 2008, who compare four different CV elicitation methods). The payment card method is regarded as efficient, robust and reliable (e.g. Bateman *et al.*, 2002; Pearce and Ozdemiroglu, 2002) and was considered the most appropriate WTP elicitation method for the purpose.
>
> After reminding respondents that they have a limited household budget and that additional money spent on supporting the chicken legislation may mean that they have less money to spend on other things, respondents were presented with a range of seven bids, of ascending value, from '5 pence per week extra' to 'more than £4 per week extra'. These bids were expressed as an additional amount that respondents would pay per week, either in the form of an increase in the price of chicken meat, or taxation for those relative few who did not consume chicken.[2] The range of amounts chosen was based on findings of pre-testing of the questionnaire and was later confirmed to be appropriate in the pilot. Respondents were asked to tick 'Yes' if they would be willing to pay each amount or 'No' if they would not. Respondents were asked to state their WTP to support the directive and its associated provisions, taken as a whole.
>
> A sample of 665 respondents was achieved. The sample contained 55% males and 45% females, with respondents living in households of 2.2 people, on average, of whom 0.4 were under 16 years of age. Average household income was £33,500. Fifty-nine per cent of respondents reported living in rural settings, i.e. in either villages or provincial towns, with 41% stating that they lived in urban areas. Almost 95% of respondents reported being consumers of chicken. Just under half of those who did not consume chicken stated their reason as being vegetarian. Respondents consuming chicken reported eating an average of 3.05 meals per week containing chicken (mode 1–2; includes takeaways and meals out), with a range from 1 to 15. Respondents reported spending a modal value of £5–10 per week (mean £7.80) on chicken purchases for their household, excluding takeaway meals and meals eaten outside the home (respondents were asked to select from categories for their response, which ranged from zero to more than £20 per week). Respondents were asked to rank their concern for farmed animals against a number of other widely held concerns, such as those in relation to the environment. Chicken eaters are most concerned about food safety, with animal welfare concerns near to last. Non-chicken eaters place food safety concerns last, but, again, animal welfare concerns are low in the order of priorities ('healthy diet' was ranked first). Respondents were asked to rank their level of concern for the welfare of broiler chickens reared in the UK on a scale of 0–10, where 0 = not concerned at all and 10 = very concerned. The sample average rank was 8.2, with chicken eaters scoring insignificantly higher than non-chicken eaters, i.e. 8.6 compared to 8.2.

Continued

Continued.

Respondents were asked a number of debriefing questions to explore the reasons for their stated WTP, and test the rationality of responses. These questions presented a number of propositional statements and respondents had to express their level of agreement with each, using an 11-point scale where 0 = 'does not reflect my views at all' and 10 = 'reflects my views completely'. Respondents were first asked to indicate whether they understood the information provided in the questionnaire about the broiler legislation. The mean sample rank score was 8.7, indicating a high level of agreement. When asked whether they understood the WTP question, the great majority of respondents indicated that they did (mean score 8.9). Respondents generally believed that the welfare of chickens reared for meat needed to be improved (8.4) and also expressed a belief that the directive itself would act to improve the welfare of chickens reared for meat (8.1). There was a relatively high level of agreement by respondents that their WTP values reflected their concerns about the welfare of broiler chickens (7.6). Overall, respondents tended to agree with the statement that they should not have to pay more to improve chicken welfare (6.6).

Respondents were asked to indicate the maximum sum that they would be willing to pay per week as an additional amount on the cost of their chicken meat purchases (or, in the case of non-chicken eaters, as an increase in taxation) to support the broiler legislation. Figure 4.1 shows the distribution of WTP responses.

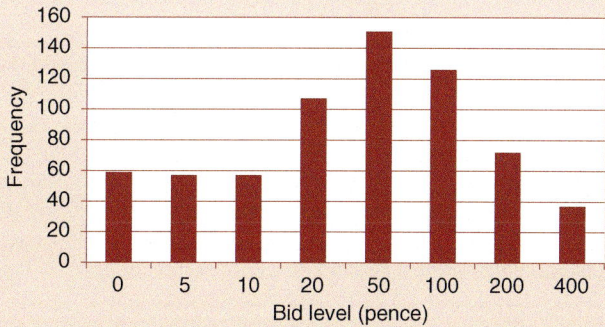

Fig. 4.1. Distribution of sample over bid acceptance levels (pence per week).

Results showed that income, membership of an organization concerned with the welfare of animals (labelled RSPCA) and having children under 16 in the household all have a significant influence on WTP. The mean and median WTPs are calculated from the intercept and variance of the log-normal regression, and can be interpreted as estimates at the sample mean of the descriptors (Table 4.1).

Table 4.1. The interval regression results for a log-normal regression estimation model (see Bennett *et al.*, 2019 for the full method and analysis).

	Estimate	SD Est	Pseudo-t
Log-normal			
Intercept	3.33	0.05	65.17**
Income	0.29	0.08	3.58**
Age	−0.01	0.14	−0.07

Continued

Continued.

	Estimate	SD Est	Pseudo-t
Female	0.19	0.10	1.82*
RSPCA[a]	0.54	0.14	3.80**
Have children	0.29	0.13	2.14**
BTSI[b]	0.08	0.10	0.86
Error variance	1.65	0.10	
Median WTP	27.97	1.43	
Mean WTP	64.11	4.52	

[a]RSPCA represents those respondents who belonged to an organization concerned with animal welfare.
[b]BTSI denotes use of a method within the questionnaire to encourage truthful reporting by respondents.
[c]Asterisks * and ** denote 2-tailed significance at the 10% and 5% significance, respectively.
[d]Number of observations=665

Median WTP is estimated at around £28 compared to a mean estimate of £64. Differences between these values reflect the highly diffuse upper tail in the log-normal distribution. Thus the median estimate is considered the better measure of central tendency, as a measure of the consumer/citizen surplus associated with the legislation.[3] This estimate was further adjusted for bias in relation to the characteristics of the sample compared with the general population and for potential 'yea saying' among responders (by giving a zero WTP to any respondent who indicated in the questionnaire that they definitely or probably would not be willing to pay something for the legislation but who then subsequently said yes to a positive WTP amount), giving an adjusted median WTP of £21.50 per household per year.

This provides an estimated benefit of the legislation to citizens of over £503 million per year (calculated by multiplying the adjusted median WTP of £21.50 per household per year by the number of households in England and Wales), equivalent to 5.3% of current consumer expenditure on chicken. This compares to an estimated £22 million per year cost of producers' compliance and government enforcement associated with the legislation, estimated from data collected from a survey of broiler producers. In conclusion, the analysis shows that people in England and Wales have a substantial stated WTP to support legislation to monitor and improve the welfare of broiler chickens in the UK. At a societal level, the benefits of the legislation, as measured by people's estimated WTP, greatly outweighed the costs. This suggests both that the current broiler legislation was worth implementing and that it is worth continuing.

and inefficient, with the potential benefits to society of interventions to improve animal welfare going unrecognized.

The second case study presented here describes an alternative method to lessen the disadvantages mentioned above.

Case Study 2 shows how a single welfare score, based on the Welfare Quality® Index, can be used to elicit people's valuations of changes in the level of welfare of different animal species as a result of policy and husbandry changes that affect welfare. The simple scoring system presented to survey participants was accepted by them as credible and was clearly understood by

Case Study 2: A welfare score for each species/production system

The approach taken in this case study is to use a framework for measuring animal welfare changes based on the Welfare Quality® Index (Welfare Quality, 2011), which then enables citizens to value specified quantitative changes in the welfare status of animals by using a stated preference choice experiment method of valuation. The measure of animal welfare is represented as a quantitative score (from 0 to 100) both with and without an intervention (all other things being equal) to clearly communicate to citizens the change in welfare status of the animals involved due to the intervention and on which they can base their willingness to pay for greater levels of welfare.[4] On the hypothetical scale, zero represents the lowest possible welfare of the animals concerned (i.e. extreme suffering) while a score of 100 denotes the highest welfare that could possibly be achieved (i.e. with all the needs of an animal met). The current Welfare Quality® system of measurement aggregates scores into four principal areas (good feed; good housing; good health; and appropriate behaviour), which are then used to assign the welfare status of animals into one of four different classes (excellent, enhanced, acceptable or not classified). Each animal production system has a different protocol for assessing welfare and for assigning scores.

The welfare scoring system presented here can be related to the Welfare Quality® Index as summarized in Table 4.2. Welfare Class 0 (i.e. 'unclassified') is represented by a welfare score of less than 40, where 40 is defined as representing the level achieved by compliance with legal minimum standards for welfare. Thus a score of 40–60 represents Welfare Class 1. Welfare Class 2 is 61–80 and Welfare Class 3 is 81–100. The method allows for the bands for each welfare class to be defined differently to that used above but, clearly, whatever system is used, it needs to be consistent and provide a reasonable reflection of the relative levels of animal welfare represented by each welfare class.

Table 4.2. Relationship between the Welfare Quality® welfare classes and the welfare scoring system.

Welfare Quality® Index	Welfare score
Welfare class 3	81–100
Welfare class 2	61–80
Welfare class 1	40–60
Welfare class 0	0–39

The study described here uses the above system to elicit people's WTP associated with a range of welfare changes (denoted by the changes in scores) across three different animal production systems – beef, pig and poultry (broiler) production.

A choice experiment survey was administered to elicit people's WTP for a range of changes in animal welfare score for the three production systems. A questionnaire was designed with the following main elements: questions about people's attitudes, opinions and consumption of meat; information about the welfare score; the choice exercise; and debriefing and socio-economic questions.

The questionnaire was pre-tested both in a focus group and in a number of personal interviews and then piloted on a sample of 50 respondents. Open-ended willingness to pay levels were also elicited from a sample of shoppers to help define the WTP bid amounts used in the choice experiment (CE) exercise. A random sample of citizens in Great Britain was used, stratified according to socio-economic group. These were then telephoned to recruit them to the survey. Those who agreed to take part were then either interviewed im-

Continued

Continued.

mediately if they had access to the internet or an interview was arranged for a few days later. Information about the welfare scoring system and the CE exercise (including provision of the choice sets) was made available on the internet for those with access or sent by post prior to the interview. Around 300 respondents participated in the survey. Participants in the survey were told that:

Animal welfare scientists and veterinarians now have developed a system for measuring the welfare of individual animals that takes account of the varying needs of different species, ages etc. The system scores the extent to which the needs and wants of the animal are met and results in an overall score on a scale of 0–100, which accurately represents the welfare of the animal in terms of its freedom from hunger, thirst, discomfort, pain, injury, disease, fear and distress, and the extent to which the animal can express natural behaviours and has a happy and contented life. A score of zero would denote extreme suffering whereas a score of 100 would denote the highest level of welfare that could possibly be achieved. The system applies over the entire life of the animal from birth to slaughter and involves regular independent monitoring of the animal's welfare throughout its life.

Participants were also told:

Assume that in your usual food store there is a section that sells meat and meat products with high welfare scores. The farmers who supply this meat will be monitored by the RSPCA. If you buy meat with a welfare score above the legal minimum (40), your weekly expenditure will rise. We now ask you to make six choices.

Respondents were then given some additional guidance before being presented with six different choice sets. This guidance included a reminder to them that their budget is limited and that more money spent on meat may mean less money to spend on other things. An example choice set is shown in Fig. 4.2.

Respondents were asked to choose one alternative only out of A, B or C (status quo) or could choose 'Don't know'. After completing the choice experiment they were asked which attributes they considered when making their choices. The choice sets contained four attributes: (i) welfare score for beef cattle; (ii) welfare score for chickens; (iii) welfare score for pigs (to explore people's WTP across farmed species), with score levels of 40, 50, 60, 70 and 90 used; and (iv) a price attribute which was their additional expenditure on meat (at £0, £6, £8, £12, £16 and £24 per month extra). Meat expenditure was therefore used as the 'payment vehicle' to obtain people's WTP for welfare changes of each of the three species.

WTP was derived from the choice experiment data using the mixed logit model with a Bayesian inference estimation procedure. The mixed logit model is a highly flexible model that can capture preference heterogeneity and approximate any random utility choice model to any degree of accuracy through appropriate specifications of the marginal utility coefficients (McFadden and Train, 2000). After testing different transformations, the preferred model that achieved the highest log-marginal likelihood specified the marginal utility coefficients of the species attributes to be normally distributed and the cost coefficient to be fixed. The Bayesian procedure for estimating the mixed logit model was carried out as described by Train (2003). Estimated coefficients derived from the mixed logit model were then used to compute WTP for a one-point increase in welfare by scaling (i.e. dividing) each welfare coefficient by the price coefficient.

Continued

Continued.

The results of the survey and analyses found that average weekly meat expenditure was £17.65 with 86% of respondents eating chicken, 68% beef, 54% pork and 55% other meat. Less than 5% said they did not consume meat. Thirty-eight per cent of respondents felt well informed about the way in which farm animals are treated and over 72% were concerned about the way farm animals are treated. In terms of attitudes and beliefs, 96% of respondents agreed that we have a moral obligation to safeguard the welfare of animals. Eighty-one per cent thought that meat from animals with high welfare has better food safety, 78% that it was healthier, 71% that it had better nutritional value and 69% that it tasted better, while 79% thought it was also better for the environment.

Suppose you could only choose one from the three baskets below. Which basket would you choose?		
Please circle only one.		Welfare scores of your **meat**
Basket A	Increase in monthly meat expenditure £ 12.00 (£ 144 per year)	 50 90 60
Basket B	Increase in monthly meat expenditure £ 16.00 (£ 192 per year)	 90 60 70
Basket C	Increase in monthly meat expenditure £ 0.00 (£ 0.00 per year)	 40 40 40
or **Don't know.**		

Fig. 4.2. Example choice set (meat expenditure).

Table 4.3 shows people's estimated mean WTP for a one-point increase in the welfare score for each of the three meat species.

It can be seen that WTP is highest for cattle welfare followed by chicken welfare, although all three WTP estimates are of similar magnitudes. The standard deviations show

Continued

Continued.

substantial variation in individual WTP, which was positively correlated to their consumption of the three different meats.

Table 4.3. Mean WTP for a one-point increase in welfare (£).

	Increase in monthly meat expenditure	Increase in annual meat expenditure
Beef/cattle welfare	0.437 (0.652)*	5.24
Pork/pig welfare	0.381 (0.479)	4.57
Chicken/chicken welfare	0.425 (0.527)	5.10

*Numbers in parentheses are standard deviations. All of the estimates are significantly different from zero at less than the 5% level of statistical significance ($p<0.05$).

The WTP estimates obtained from using the method outlined in this paper seem credible in terms of their orders of magnitude. For example, across the three meats, people were willing to pay an average of 16% increase in meat expenditure for a ten-point increase in welfare score. Moreover, a separate survey (with an entirely separate sample, not reported here) was undertaken at the same time as the one presented here, which also used the welfare scoring system but which used the contingent valuation (CV) method instead of the choice experiment (CE) method and elicited values of similar orders of magnitude. Thus, over the 40–80 welfare score range, the CE survey results gave a £147/year/ten-point increase in welfare score for the three meats together while the CV survey results gave £114/year/ten-points.

them (e.g. as demonstrated in focus group discussions and survey respondents' comments in the open-ended debriefing question). The advantage of using a single welfare score to describe the welfare status of animals is that we can then obtain people's valuation of defined changes in animals' welfare rather than their valuation of a husbandry or other system change that might impact on welfare as obtained by previous animal welfare valuation studies. Assuming that this welfare valuation is transferable, regardless of the husbandry or other means by which a change in welfare status is brought about, then policy makers no longer need to commission repeated valuation studies for their impact assessments every time there is a (proposed) change in animal production practices (e.g. requiring a change in legislation).

Of course, potential problems remain as to both the ethical underpinnings of this simple system and its practical application. These include (i) people may value (and have a WTP for) the means by which welfare is improved as well as the welfare outcomes (e.g. people often like animals to have access to a natural environment rather than being contained solely within indoor environments even though the former could compromise welfare due to possible increased risk of disease or predation or inclement weather); (ii) there are practical and ethical considerations underlying the scoring of animal welfare (see Veissier *et al.*, 2011); (iii) people's willingness to pay as elicited in the example above is not solely for animal welfare *per se* but includes their WTP for a range of attributes that they perceive to be associated with higher levels of animal welfare such as food quality, including taste, food safety and environmental credentials; (iv) the sample size of the above study (300) was relatively small as this was a 'proof of concept' study; a larger-scale exercise would need to be undertaken to provide estimates usable in a policy context; (v) WTP estimates will change over time and so will need to be re-estimated/updated periodically (and this would be useful validation showing the consistency or otherwise of both the method and of people's WTP); and (vi) the valuations presented here are 'private benefits' relating to personal consumption and not social ones (see Bennett, 1995). People's WTP for meat from animals with higher welfare does not take into account the benefits to others in society that accrue from these others, knowing that people who consume higher-welfare meat are promoting good animal welfare. This is a 'positive externality' of private consumption decisions (these externalities can be negative also, for example where people's consumption is perceived as promoting animal suffering) and takes the form of a 'public good' where all of those in society who care about the welfare of animals (the vast majority in the UK as shown by this and previous surveys) experience a benefit (or cost) associated with the behaviour of others that they perceive to be resulting in good or bad animal welfare.

In addition, there is a significant challenge for animal welfare scientists to be able to provide both *ex ante* and *ex post* welfare assessments of an intervention or policy. Clear protocols are necessary for panels of animal welfare scientists to be able to derive the appropriate welfare scores to assess changes to welfare as a result of specified interventions. These scores can only ever be imperfect human assessments of welfare impacts on animals and may or may not accord well with animals' preferences. Animal welfare science has come a long way over the last 20 years in terms of both fundamental knowledge, for example in relation to better understandings of animal sentience and cognition, and in terms of methodological approaches to measuring animal welfare. However, where an *ex ante* assessment of the impact of a particular intervention on animal welfare is involved, then there may be limited evidence (e.g. from a small number of experiments or small-scale trials) as to the impacts on the welfare of animals affected in, for example, larger-scale (perhaps commercial) settings.

4.6 Implications of Preferences for Society and for Policy

Despite the scientific developments in our understanding and measurement of animal preferences and animal sentience, there is no direct means for incorporating these into a societal welfare function. However, this greater understanding, if widely and effectively communicated, will change people's perceptions, empathy and preferences with respect to the welfare of animals and its importance/prioritization in relation to other considerations. This change in human preferences will become increasingly evident in our market transactions (subject to the provision of appropriate information on animal welfare provenance of products via labelling, assurance schemes, etc.), regulation and lobbying (the latter largely through NGOs concerned with animal welfare).

Indeed, the paucity of appropriate information on animal welfare in society has been a major impediment to the improvement of the wellbeing of animals. Welfare considerations remain somewhat shrouded in mystery for members of the general public, the vast majority of whom would identify themselves as 'animal lovers'. When asked, the vast majority of citizens in the UK and many other countries, would say that they are concerned about the welfare of animals and think it should be improved. But their perceptions and understanding of welfare are limited, very partial and poorly, if not wrongly, informed. Understanding is sometimes based on limited and partial media coverage, comments and views of family and friends, etc. This is true in relation to food-producing animals as well as others. Food consumers have little knowledge of animal production systems and husbandry, the conditions under which animals are kept or their welfare status. Information about the welfare of animals used to produce our food available to consumers is also very limited, partial and sometimes misunderstood by consumers. There is a need for much better information on welfare to be made available to consumers, and to citizens more widely, for them to be able to make rational choices in relation to the welfare attributes associated with animal products. In 2006, the Farm Animal Welfare Council in the UK produced a report on welfare labelling which recommended the adoption of an EU-wide, single, accredited, mandatory animal welfare labelling system applied to all animal-based products (FAWC, 2006). Such a system has yet to be introduced. In the 2015 Eurobarometer survey of nearly 28,000 citizens across 28 countries, 47% did not believe there is currently a sufficient choice of animal welfare-friendly food products in shops and supermarkets and only 10% certainly believed that there was (European Commission, 2016a). Improvement in the information available to citizens would greatly help individuals to better formulate and express their preferences in relation to animal welfare. The issue of welfare labelling is discussed in more detail in Chapter 3.

The practical scientific assessment of the welfare status of animals (both as individuals and groups) has also greatly developed over the last 20 years. Butterworth *et al.* (2018) provide a useful introduction to the challenges, issues and current thinking in relation to welfare assessment. One practical outcome of this scientific development has been the Welfare Quality assessment protocols for a range of farm animal production systems (see www.welfarequality.net/en-us/reports/assessment-protocols). This advancement can provide a much better understanding of the welfare impacts on animals of a range of activities and interventions both current and prospective, which can be used to help guide policy in relation to animal welfare, how we use and interact with animals and our duty of care towards them.

Given the inability of animals to express their preferences and to be included within the utilitarian metric of a societal welfare function, there is a clear need for welfare advocacy on behalf of animals. This is a role currently played by government and by a number of animal welfare charities on society's behalf, as guardians of the welfare of animals. This stewardship/

custodianship role is essential if the interests of animals are to be given consideration (not necessarily an 'equal consideration of interests'). Of course, human preferences for animal welfare remain a key determinant of the nature and extent of this stewardship/custodianship. This stewardship must be informed by the scientific evidence on animal sentience and preferences, and on measurement of the welfare states of animals.

This means that society must rely on a pragmatic (rights-based) approach for people (e.g. legislators) to act on animals' behalf and, taking animal preferences into account together with evidence-based knowledge of impacts on animal welfare, put in place safeguards via regulation and other policy instruments[5] which protect animal welfare, prevent cruelty, promote positive welfare, give animals certain rights not to be treated inhumanely and for people to have a duty of care towards animals. To do this comprehensively, is likely to require a change in governance.

For example, there have been calls for an Animal Welfare Protection Agency, which has the specific task of safeguarding the welfare (and rights) of animals. It would need to do this both on behalf of human sentiments towards animals (i.e. the effects on human utility/welfare due to human empathy with animals and concerns for their welfare) and in relation to the preferences and welfare of animals *per se*. Such a body would need extensive and significant powers given the wide-ranging nature of impacts on animal welfare across human activities. In the UK, organizations which have taken on some of this role include various charities (CIWF, PETA, RSPCA, SSPCA, RSPB), governmental bodies (FAWC, Home Office, Defra) and others such as the British Veterinary Association (which has an Animal Welfare Foundation). Indeed, the charity sector has perhaps been most visible in this regard and, arguably, very impactful. There are numerous animal welfare advocacy groups in Europe including the Eurogroup for Animals. In the USA, the Humane Society of the USA has an important role in protecting animal welfare together with organizations such as the American Society for the Prevention of Cruelty to Animals and Animal People. Internationally, the World Organisation for Animal Health (OIE) is an intergovernmental body in charge of improving animal health (and welfare) worldwide and has some 174 member countries/territories. The United Nations' Food and Agriculture Organization and the European Food Safety Authority also have an interest in improving animal welfare while the concept of animal welfare is enshrined in Article 13 of the Treaty on the Functioning of the European Union, which recognizes animals as sentient beings (European Commission, 2016b). In addition, there are major international charities such as the International Fund for Animal Welfare and World Animal Protection which have a global presence. Arguably, the wider the sphere of influence of an organization, the greater the compromises that need to be made in terms of setting standards for animal welfare and care. This means that national animal welfare protection agencies are likely to be able, and perhaps required, to achieve higher levels of animal welfare in certain countries rather than relying on the standards of international bodies agreed by many Member States.

There are substantial benefits to society of people being able to better formulate and express their preferences in relation to animal welfare. The first is the added value that both consumers and producers may derive from the consumption and production of higher welfare animal products which are seen as higher quality in relation to a number of product attributes. Second, there is the reduced cognitive dissonance of consumption and production of animal products with relatively low welfare standards for those with empathy for animals (which is most consumers/producers) and the satisfaction that we can derive from knowing that animals are well cared for and have high levels of welfare. A third category of benefits relates to knowing we are 'doing the right thing' as moral agents and guardians of animals. Finally, there is a broader benefit related to a caring society. In its Global Animal Welfare Strategy, the World Organisation for Animal Health (OIE Global Animal Welfare Strategy, 2017) acknowledges that 'animal welfare is closely linked to animal health, the health and well-being of people, and the sustainability of socio-economic and ecological systems'. Moreover, our preferences in relation to animal welfare, and how we act on them, say something about us as individuals and as a society. US Congress member Jim Moran (2014) sums this up well:

> [T]reating animals inhumanely undermines the moral standing of society. Disregarding the health and welfare of animals arguably makes it easier for people to disregard the health and welfare of fellow humans. It limits our capacity for empathy and constrains our ability to look beyond self-interest. The treatment of animals can be a catalyst for a more just and compassionate society, or it can be a symptom of a society that has lost its moral compass.

People need to be able to express their preferences regarding the treatment of animals but this is greatly constrained by available information and a paucity of accessible choice alternatives. If direct preference expression by people in relation to animal welfare, for example through markets, is inhibited, then government or an alternative body acting on behalf of society must take up the role to represent societal animal welfare preferences and act as custodians to protect the welfare of animals. Failure to act on people's preferences in relation to animal welfare will mean that human society (and ecological systems) suffers the consequences, as well as animals.

Notes

¹ The term 'animal welfare' is used here in the context of the World Organisation for Animal Health definition (OIE Terrestrial Animal Health Code, 2018) that 'Animal welfare means the physical and mental state of an animal in relation to the conditions in which it lives and dies. An animal experiences good welfare if the animal is healthy, comfortable, well nourished, safe, is not suffering from unpleasant states such as pain, fear and distress, and is able to express behaviours that are important for its physical and mental state.'

² A zero amount was not presented to respondents at this point because (i) respondents could answer 'No' to all the bids presented to them including the lowest 5 pence per week bid and (ii) a previous question had already asked respondents whether they would be likely to be

willing to pay something for the legislation on a five-point scale ('Definitely yes', 'Possibly yes', 'Not sure', 'Probably not', 'Definitely not').

[3] Note that the very few in the sample (34 of 665) who did not eat chicken had an average WTP of around twice that of chicken eaters.

[4] In principle, people could be asked their 'willingness to accept' a reduction in animal welfare but this introduces additional issues related to willingness to accept values (see Whittington *et al.*, 2017).

[5] Bennett and Appleby (2010); Bennett and Appleby (2011); Bennett and Thompson (2018); and FAWC (2008) lay out the range of policy instruments that can be used to improve animal welfare. Institutional change is an important category of policy instrument, as is legislation.

References

Bateman, I.J., Carson, R.T., Day, B., Hanemann, M., Hanley, N. *et al.* (2002) *Economic Valuation with Stated Preference Techniques: A Manual.* Edward Elgar, Cheltenham, UK.

Bekoff, M. (2009) *Encyclopedia of Animal Rights and Animal Welfare*, 2nd edn. Vol. 569. Greenwood Press, Westport, Connecticut.

Bennett, R. (1995) The value of farm animal welfare. *Journal of Agricultural Economics* 46, 46–60.

Bennett, R., Balcombe, K., Jones, P. and Butterworth, A. (2019) The benefits of farm animal welfare legislation: the case of the EU broiler Directive and truthful reporting. *Journal of Agricultural Economics* 70(1), 135–152. DOI: 10.1111/1477-9552.12278.

Bennett, R. and Appleby, M. (2010) Animal welfare in the European Union. In: Oskam, A., Meester, G. and Silvis, H. (eds) *EU Policy for Agriculture, Food and Rural Areas*. Wageningen Academic Publishers, Wageningen, The Netherlands, pp. 243–251.

Bennett, R. and Appleby, M. (2011) Animal welfare policy in the European Union. In: Oskam, A., Meester, G. and Silvis, H. (eds) *EU Policy for Agriculture, Food and Rural Areas*, 2nd edn. Wageningen Academic Publishers, Wageningen, The Netherlands, pp. 249–258.

Bennett, R., Butterworth, A., Jones, P., Kehlbacher, A. and Tranter, R. (2011) *Valuation of Animal Welfare Benefits*. Report to the Department for the Environment, Food and Rural Affairs, University of Reading, Reading, UK.

Bennett, R., Kehlbacher, A. and Balcombe, K. (2012) A method for the economic valuation of animal welfare benefits using a single welfare score. *Animal Welfare* 21(1), 125–130. DOI: 1 0.7120/096272812X13345905674006.

Bennett, R. and Thompson, P. (2018) Economics. In: Appleby, M.C., Olsson, I.A.S. and Galindo, F.. (eds) *Animal Welfare*, 3rd edn. CAB International, Wallingford, UK, pp. 335–348.

Bentham, J. (1789) *Introduction to the Principles of Morals and Legislation. 1996 Imprint.* Clarendon Press, Oxford, UK.

Blackorby, C. and Donaldson, D. (1992) Pigs and guinea pigs: a note on the ethics of animal exploitation. *The Economic Journal* 102, 1345–1369.

Butterworth, A., Mench, J.A., Wielebnowski, N. and Olsson, I.A. *et al.* (2018) Practical strategies to assess (and improve) welfare. In: Appleby, M.C., Olsson, I.A.S. and Galindo, F.. (eds) *Animal Welfare*, 3rd edn. CAB International, Wallingford, UK, pp. 335–348.

Dawkins, M.S. (1990) From an animal's point of view: motivation, fitness, and animal welfare. *Behavioral and Brain Sciences* 13(1), 1–9. DOI: 10.1017/S0140525X00077104.

Defra (2010) Study to evaluate the effectiveness of Regulation (Directive 2007/43/EC) in England and Wales. EVID4 Evidence Project Final Report (Rev. 06/11) (2017). Defra Council Directive 2007/43/EC. Laying down minimum rules for the protection of chickens kept for meat production, Impact Assessment No Defra. Defra, London.

European Commission (2016a) *Eurobarometer 442. Attitudes of Europeans Towards Animal Welfare.* European Commission, Brussels.

European Commission (2016b) Animal welfare. Available at: http://ec.europa.eu/food/animals/welfare/index_en.htm

European Commission Council Directive 2007/43/EC (2007) Laying down minimum rules for the protection of chickens kept for meat production. Official Journal of the European Union, L 182/19. Available at: http://eur-lex.europa.eu/legal-content/EN/TXT/PDF/?uri=CELEX:32007L0043&rid (accessed 12 July 2007).

FAO (2001) Guidelines for humane handling, transport and slaughter of livestock. Available at: http://www.fao.org/3/X6909E/x6909e00.htm#Contents (accessed 13 January 2020).

FAWC (2006) *Report on Welfare Labelling*. Farm Animal Welfare Council, London.

FAWC (2008) *Opinion on Policy Instruments for Protecting and Improving Farm Animal Welfare*. Farm Animal Welfare Council, London.

Fraser, D. and Nicol, C.J. (2018) Preference and motivation testing. In: Appleby, M.C., Olsson, I.A.S. and Galindo. (eds) *Animal Welfare*, 3rd edn. CAB International, Wallingford, UK, pp. 335–348.

Heinzen, R.R. and Bridges, J.F.P. (2008) Comparison of four contingent valuation methods to estimate the economic value of a pneumococcal vaccine in Bangladesh. *International Journal of Technology Assessment in Health Care* 24(4), 481–487. DOI: 10.1017/S026646230808063X.

Hursh, S.R. (1980) Economic concepts for the analysis of behaviour. *Journal of the Experimental Analysis of Behaviour* 34, 219–238.

Johansson-Stenman, O. (2018) Animal welfare and social decisions: is it time to take Bentham seriously? *Ecological Economics* 145, 90–103.

Lagerkvist, C.J. and Hess, S. (2011) A meta-analysis of consumer willingness to pay for farm animal welfare. *European Review of Agricultural Economics* 38(1), 55–78. DOI: 10.1093/erae/jbq043.

Loomis, J.B. (2014) Strategies for overcoming hypothetical bias in stated preference surveys. *Journal 670 of Agricultural and Resource Economics* 39(1), 34–46.

Louviere, J.L., Hensher, D.A. and Swait, J.D. (2000) *Stated Choice Methods*. Cambridge University Press, Cambridge, UK.

Marshall, A. (1947) *Principles of Economics*, Reprint of 8th edn. Macmillan and Co. Ltd, London.

McFadden, D. and Train, K.E. (2000) Mixed MNL models for discrete response. *Journal of Applied Econometrics* 15, 447–470.

McInerney, J. (1994) Animal welfare: an economic perspective. In: Bennett, R.M. (ed.) *Valuing Farm Animal Welfare*. University of Reading, UK, pp. 9–25.

Mill, J.S. (1863) *Utilitarianism*, 1st edn. Parker, Son & Bourn, West Strand, London.

Mitchell, R.C. and Carson, R.T. (1989) *Using Surveys to Value Public Goods. The Contingent Valuation 683 Method*. Resources for the Future, Washington, DC, p. 684.

Moran, J. (2014) Respect of animals would lead to more empathetic human society. The Hill, 31 July 2014.

OIE Global Animal Welfare Strategy (2017) *World Organization for Animal Health*. Paris.

OIE Terrestrial Animal Health Code (2018). Available at: www.oie.int/en/standard-setting/terrestrial-code/access-online (accessed 10 April 2019).

Pearce, D. and Ozdemiroglu, E. (2002) *Economic Valuation with Stated Preference Techniques*. Department for Transport, Local Government and the Regions, Rotherham, UK, pp. 695–696.

Regan, T. (1982) *All that Dwell Therin: Animal Rights and Environmental Ethics*. University of California Press, Berkeley and Los Angeles, California.

Singer, P. (1975) *Animal Liberation*. Jonathan Cape, London.

Smith, A. (1790) *The Theory of Moral Sentiments*, Revised Edition. T. Cadell, London. Republished in 1975 by Oxford University press, Oxford, UK.

Train, K. (2003) *Discrete Choice Methods With Simulation*. Cambridge University Press, Cambridge, UK.

Veissier, I., Jensen, K.K., Botreau, R. and Sandøe, P. (2011) Highlighting ethical decisions underlying the scoring of animal welfare in the welfare quality scheme. *Animal Welfare* 20, 89–101.

Welfare Quality (2011) Welfare quality network. Available at: www.welfarequalitynetwork.net (accessed 10 April 2019).

Whittington, D., Adamowicz, W. and Lloyd-Smith, P. (2017) Asking willingness-to-accept questions in stated preference surveys: a review and research agenda. *Annual Review of Resource Economics* 9(1), 317–336. DOI: 10.1146/annurev-resource-121416-125602.

Zhang, L. (2009) An exchange theory of money and self-esteem in decision making. *Review of General Psychology* 13(1), 66–76. DOI: 10.1037/a0014225.

5 Animal Welfare and Farm Economics: An Analysis of Costs and Benefits

JARKKO K. NIEMI*

Natural Resources Institute Finland (Luke)

Summary

Livestock farming uses resources such as land, labour, capital and animals to produce goods and services with a use value to people. Simultaneously, it also produces goods and services that have non-use values to humans. Profit maximization is a fundamental goal in farm-level economic analysis of animal welfare. For a farmer to have an incentive to adopt an animal-friendly farming practice, it is essential that additional costs and loss of revenues, if any, are smaller than possible savings in the costs plus additional revenues obtained through animal welfare improvement. This chapter reviews the principal relationships between animal welfare and costs, benefits and profit associated with farming. It also discusses the implications that animal welfare improvements can have for farm productivity, efficiency and profits, and the externalities that welfare improvements can lead to. Different cases leading to animal welfare improvements are examined. These include: (i) the case where animal welfare improvements are motivated by changes in farm productivity, for instance due to reduced disease incidence; and (ii) the event that an animal welfare improvement is motivated by additional price premiums obtainable from the market.

5.1 Introduction

Solving a monetary optimum provides useful information for decision makers. It can highlight the roles of opportunity cost and cost competitiveness, which is important for the continuation of a livestock business, and evaluate the effect of policies on farming. Studies suggest that besides maximizing profits, farmers

---

*jarkko.niemi@luke.fi

can also have other objectives. Hansson and Lagerkvist (2015) identified use and non-use values that underlie dairy farmers' decision making with respect to animal welfare. The relevant use values were related to the farmer being able to continue the business, earn a living from the business and not being tied to the farm (i.e. having time available for other things), product quality and work environment. Identified non-use values were related to avoidance of suffering, being able to further improve the welfare of animals, the farmer feeling good him/herself, ethical considerations, a feeling of doing 'the right thing', and animals eating properly (i.e. functioning as animals should). This highlights the view that both monetary and non-monetary factors can have a value, that they both can influence farmers' choices regarding animal welfare, and both may need to be addressed in economic analyses (see Chapters 3 and 4 of this book for discussion on demand-side issues). These objectives can be taken into account in profit maximization by imposing constraints to the maximization problem of the farmer and assigning values to inputs and outputs other than the monetary revenues or costs.

This chapter reviews the principal relationships between animal welfare and the costs, benefits and profit associated with farming. Section 5.2 reviews different types of cost and benefit factors related to animal welfare; Section 5.3 examines profit functions and farmers' decision making; further sections provide empirical examples on the costs and benefits of animal welfare.

5.2   What Costs and Benefits Are There Related to Animal Welfare?

Changes in animal welfare can influence farm economy in many ways. When analysing costs and benefits of farm animal welfare, the main focus is on how animal welfare, or interventions to improve it, increases production costs and whether production and revenues are forgone; and comparing these with additional revenues plus saved costs. Currently, there are farm animal welfare concerns related to factors such as space allowance and confinement practices in livestock farming, floor materials and comfort around resting (e.g. floor and bedding materials), access to pasture and outdoors, animals' opportunities to express their natural behaviours, and painful mutilations such as castration, tail docking or dehorning (and whether pain relief is provided, see, e.g. Grethe, 2017; Wallenius et al., 2020). Interventions that improve animal welfare and address these and other concerns require different types of inputs, and input usage incurs cost (e.g. Bennett, 2012; Jones et al., 2018).

The profit of the farm is defined as the revenues obtained from farming minus variable costs and fixed costs. *Variable costs* are incurred by using inputs which can be controlled by the livestock production system and these costs can be adjusted in the short term. The use of these inputs can vary over time and can be adjusted case by case. Feed, veterinary care and working time used to care for the animals are examples of variable inputs. Practical examples include: taking better care of animals by monitoring and interacting with them requires labour; providing prevenatitive health care incurs veterinary cost; improvements in feeding may increase feed costs if higher-quality nutrition requires costly feed materials that are better suited for animals; providing enrichments and nest-building material for animals to express their natural behaviour requires materials and working time; and providing cows with access to pasture incurs labour costs and materials needed to establish and maintain pastures and fences. While all these costs contribute directly to animal productivity and farm net revenues, it is likely that increased welfare levels will exceed profit-maximizing levels, especially if more profitable uses of limited capital are available, i.e. animal welfare often carries an opportunity cost at farm level.

Fixed inputs can have a major impact on animal welfare. For example, switching from tied stalls to loose housing dairy production requires costly adjustments in housing, and using floor materials which are friendly to the hoof are costly investments and influence over several production cycles. Although fixed costs cannot be adjusted in the short term, the usage rate of fixed inputs can influence farm revenues. For example, housing costs can be the same irrespective of whether 90 fattening bulls or 100 bulls are kept in the given animal shelter as long as these 100 animals can be legally placed in the shelter, and irrespective of how they are being treated. However, reducing stocking density of fattening bulls by 10% implies that fewer animals will be sold to slaughter. Although fixed costs will remain at the same level, it implies that fixed costs per animal will rise by about 10%. Hence, more revenue must be received per animal to cover the costs of fixed inputs. To put it another way, a farm's fixed costs do not increase even if production is increased by 10%, but variable costs and revenues will increase. The potential of animal welfare improvements to generate additional revenue is therefore important when considering reduction in stocking density (e.g. Herva *et al.*, 2015).

When analysing animal welfare from an individual farm's perspective, only *private costs and benefits* to this farm are considered. While farmers may argue that the net changes in their profits are substantial, it is not the only economic consideration relevant to animal welfare. *Externalities* can be relevant in the event of animal welfare. An externality refers to a cost or benefit incurred or received by a third party, such as the animal, as a consequence of decisions taken by an economic agent (the farmer in this case). However, the third party has no control over the creation of that cost or benefit. These wider societal implications of farm animal welfare were discussed in Chapter 2.

The potential conflicts between animal welfare and efficient farming may be reduced if there are the financial benefits to firms from improving animal welfare. These benefits include: increased profits through reduced mortality, improved health, improved product quality, improved disease resistance and reduced medication, lower risk of zoonotic and food-borne diseases, improved farmer job satisfaction and labour input usage, and contributions to corporate social responsibility, which may be linked to higher product prices (see Dawkins, 2016, for discussion).

Improvements in animal welfare can provide also additional revenues. These revenues can be obtained because of increased productivity or because of additional quality characteristics that animal welfare can provide to the markets. Improvements in animal health are typical examples of changes that will increase productivity. Healthier animals will typically produce a higher yield and incur less input cost, such as veterinary medicine costs (Niemi *et al.*, 2018). However, even animal health improvements are not profitable *per se*. Their financial viability depends on the case. Some animal welfare improvements can also provide farms with access to price premiums paid for animal-friendly or high-quality production, or allow sale of the products to markets where the consumers are willing to pay a higher price for the product. Moreover, as D'Eath *et al.* (2016) illustrate, the decision whether a farmer adopts an animal welfare-improving intervention can be influenced by both the average costs and benefits, and the volatility of these.

5.3   Solution for a Profit Function Involving One Output and Multiple Inputs

In order to understand the role of costs and benefits on the farm, it is useful to look at them theoretically. Various textbooks (e.g. McFadden and Fuss, 1978; Rushton, 2009) provide an outline of production economics. Production and cost theory helps us to understand how changes in costs, revenues, profits and productivity can influence firm-level decision making. As mentioned earlier, a behavioural assumption that the farm manager maximizes the farm's profits is made. Because an individual farm or farmer is a price-taker, the farm's revenue is determined by the market prices of inputs and output(s) and by production technology, which sets limits on possible production. Hence, the farmer chooses the amount of inputs and outputs so that they are at the optimal level given the market prices and production technology. The profit of the farm (π) is determined by a profit function:

$$\pi = \max_{x} py(x_1, ..., x_n) - \sum_{i=1}^{n} w_i x_i - C$$

where p is the price of output, y is the amount of output produced, x_i is the amount of input i used, w_i is the price of input i, for $i = 1,...,n$ inputs, and C is the fixed cost.

In order to maximize the profit, and to minimize the costs, the production function must be known. The production function represents a 'technology' that is used to convert inputs to output. It describes how much output (e.g. milk) a farmer can produce with a certain amount of inputs. The algebraic solution to the profit maximization problem involves differentiating the profit function with respect to the inputs:

$$\frac{\partial \pi}{\partial x_i} = p \frac{\partial y(x_1, ..., x_n)}{\partial x_i} - w_i = 0$$

Therefore, the optimum satisfies the condition:

$$\frac{\partial y(x_1,...,x_n)}{\partial x_i} = \frac{w_i}{p}$$

The left-hand side of the expression above is also called the marginal physical product. It represents how much additional output can be obtained by using one more unit of input.

So-called duality between output and input implies that for a given amount of output the profits are maximized when also the costs are minimized. Input use which minimizes the cost of production for that amount of output also maximizes the profit if the quantity of output is fixed. Therefore the solution to the profit maximization problem is actually a two-step solution: first, the least costly way to produce output is solved and thereafter a profit-maximizing quantity of output is resolved.

Animal welfare can be treated in different ways in the profit function depending on the context of the decision problem and the nature of animal welfare in the problem at hand. First, animal welfare can be regarded as output and therefore the amount of animal welfare produced is represented by the production function. The second option is to consider animal welfare as a constraint that must be met, because, for instance, the principal purchasing the animals may have standards for the minimum level of animal welfare. Enhancing animal welfare may also incur costs but no additional revenues are generated with improved animal welfare. In that event, it is important whether enhanced animal welfare is a joint input (e.g. Leontief-technology) such that increasing the use of another input is rewarded and, simultaneously, animal welfare is also improved. Third, animal welfare can be associated with fixed inputs such as space allowance.

When animal welfare is an externality, improving it does not provide additional benefit to the farmer and therefore it does not influence the optimal level of animal welfare. As the solution above illustrates, fixed costs do not influence the decision to enhance animal welfare and neither do externalities as they do not add costs or benefits to the farmer.

Figure 5.1 illustrates the law of diminishing returns which implies that when more output is produced, each additional unit of output produces additional profit until the output is increased to the level where marginal cost is equal to the price of product. The profits are maximized where marginal (or additional) revenues are equal to marginal (or additional) costs, as at this point average cost AC* per unit of output is still below the price line, and the profit is defined as $p - AC*$ multiplied by $y*$.

Production and cost theory involves several assumptions which must be met to ensure that the analysis is robust. Rushton (2009) summarizes the important initial assumptions of production economics. The production process is assumed to be stable during the period of study. The livestock system selects the most efficient known technology which (in the static theory) is applied throughout the planning period. Production functions are also drawn as smooth, well-behaved curves and inputs are taken as being divisible and mobile. The same assumptions of homogeneity and perfect divisibility are

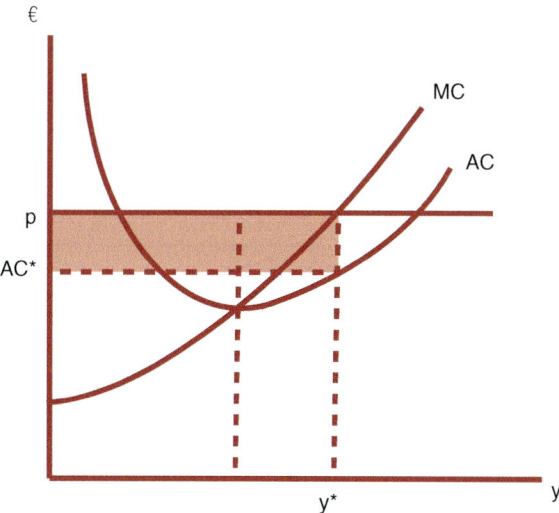

Fig. 5.1. Profit maximization, marginal and average costs.

also applied to products. The goals of the decision maker are also essential: the decision makers are assumed to be rational and to maximize profits. Specifying the goals is important because they are reflected by choices made by decision makers. In more advanced analyses, factors such as system dynamics, risk and uncertainty can be incorporated into the analysis.

A profit-maximizing firm chooses its inputs and outputs so that the maximum economic profit is achieved. In order to analyse animal welfare from a farm-economic perspective, one must pay attention to how animal welfare is associated with fixed and variable costs of farming, revenues (including effects on the price of output and possible subsidy rates) and production function. If there are price or quality premiums obtainable through investing in animal welfare, it does not violate the price-taking assumption because low and high levels of animal welfare can also be regarded as the farm choosing between producing two different products.

Some argue that 'profitable animals are happy animals', which suggests that the profit incentive is perfectly aligned with the incentive to maximize animal welfare. However, elementary production economics indicates that the level of input usage that maximizes production or yields is not the same as the level of input usage which maximizes profits. When inputs are costly, a profit-maximizing farmer will choose to produce less than is biologically possible. Similar reasoning suggests that a profit-maximizing livestock producer will choose levels of production that do not coincide with levels of animal welfare that are maximal for the animal itself (see Lusk and Norwood, 2011). Moreover, one can argue that production or yield itself is not the best measure of the level of animal welfare.

Figure 5.2 elaborates the logic of the most likely consequences of improved animal welfare in a simplified manner. Prior to the improved animal welfare, the farm operates with marginal costs MC1 and receives price P1 as the marginal value of the output, and curve 1 baseline as the profit corresponding to these costs and benefits. The four cases presenting improved animal welfare in Fig. 5.2 are:

- In the event that an animal welfare improvement would be available without any change in the production costs but a price premium P2, the farm would continue operating with marginal cost MC2 = MC1, and hence meet profit curve 2, 'price premium'. Hence the farm would receive a higher profit than in the baseline case and also have an incentive

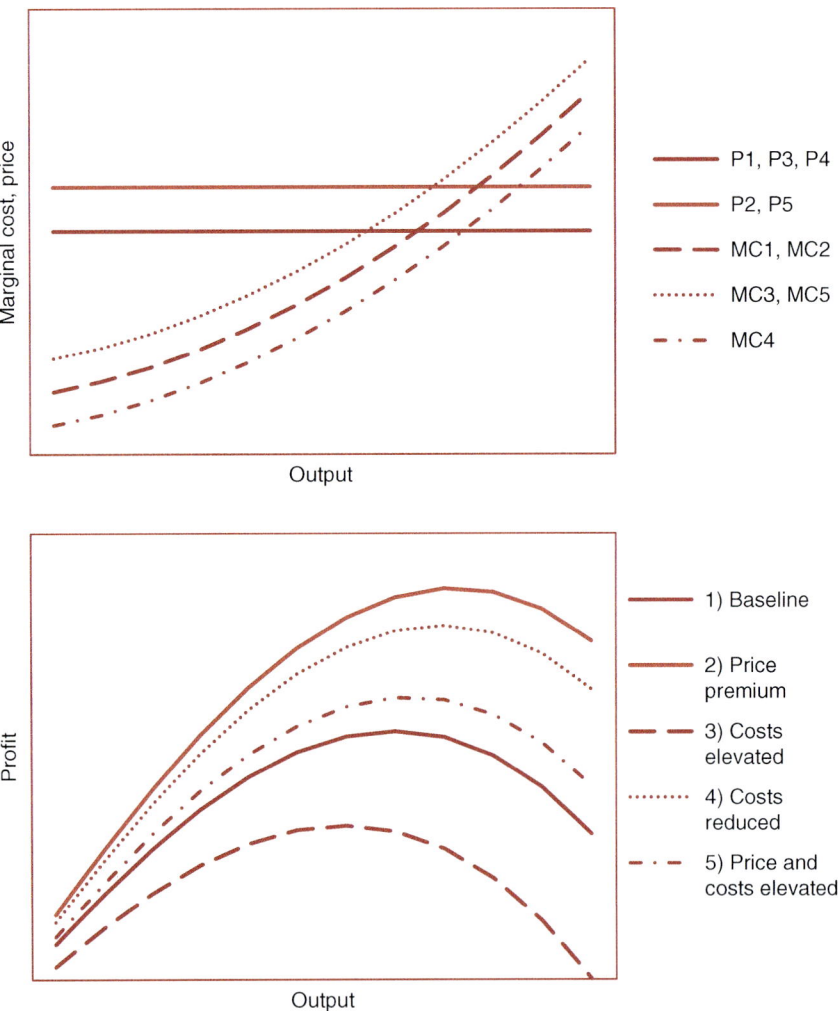

Fig. 5.2. Illustration of marginal cost (MC1,..., MC5) and revenue (P1,..., P5) (upper panel) and profit (1,..., 5) (lower panel) for situations describing the case of standard animal welfare with a standard price and costs, and four cases of improved animal welfare leading to rising product value (i.e. a price premium, P2, P5), rising production costs (MC3, MC5) or reduced production costs (MC4). Dots in the upper panel indicate the maximized profit.

to increase its production volume. In this scenario any changes in input use should be outweighed by another, opposite, change.

- In the event that an animal welfare improvement would increase production costs up to marginal cost curve MC3 without influencing the output price (P3), the farm would face profit curve 3, 'costs elevated', which provides less profit than the baseline case, and it would be optimal for a profit-maximizing farm to reduce production. This case could take place, for instance, when an elevated animal welfare standard is imposed by a newly enforced regulation.

- In the event that an animal welfare improvement would reduce production costs to marginal cost curve MC4 without influencing the output price (P4), the farm would face profit curve 4, 'costs reduced', which provides a higher profit than the baseline case, and it would be optimal for a profit-maximizing farm to increase production. This case could take place, for instance, when animal disease is reduced.

- In the event that an animal welfare improvement would increase production costs up to marginal cost curve MC5 and simultaneously provide access to a price premium leading to output price P5, the farm would face profit curve 5, 'price and costs elevated'. However, it depends on the relative change between price and marginal costs whether this case would lead to a higher or lower profit than the baseline. Case 5 could take place, for instance, upon introduction of a production system that is labelled 'animal-friendly'.

5.4   Animal Welfare and Animal Productivity

The concepts of efficiency and productivity are relevant in the context of farm-level analysis of influences of animal welfare but neither one of them, in fact, describes economic impacts of animal welfare. *Productivity* can be defined as the amount of output obtained per unit of input. Productivity can be measured as partial productivity (e.g. animal's weight gain per kg of feed) or as total factor productivity (e.g. total value of inputs used per outputs obtained). Partial productivity measures are often informative, but they can provide a misleading indication of overall productivity when considered in isolation.

McInerney (2004) presented a hypothetical relationship between increasing farm animal productivity (human benefit) and animal welfare. According to McInerney, increases in productivity arising from better nutrition, health and housing provided to farm animals compared to their natural (wild) environment can at first enhance animal welfare as well as providing human benefit. However, at higher levels of productivity, animal welfare begins to decline with increasing productivity as more intensive production systems are adopted. Legislation provides a limit to this process, setting a minimum level of animal welfare, i.e. a point beyond which further increasing livestock productivity is considered as cruelty and unlawful.

The key question is: What is the desired or appropriate level of animal welfare from the human perspective? How much human benefit from increasing productivity should be sacrificed to raise animal welfare above the current baseline?

The type and magnitude of welfare improvement affects whether it has a positive or negative impact on productivity. Positive impacts on productivity can possibly be obtained in cases where output is increased (such as improving growth, yield or value of product, or reducing disease) in relation to inputs used, or input usage is decreased in relation to the yield (increasing input intensity of production such as feed-to-gain ratio). By contrast, increasing input use without any adjustment in production is expected to reduce productivity. Figure 5.3 illustrates a specific situation where increasing the presence of three animal stressors (high temperature, restricted space allowance or regrouping of animals), one at a time or two or more at the same time, decreases total productivity. Therefore, eliminating these stressors is expected to both improve animal welfare and improve productivity. Nevertheless, this doesn't yet imply that the profitability of farming would also be improved because the full costs of eliminating the stressors are not specified.

Efficiency, by contrast, refers to what has been attained when compared to what could have been attained. Either technical efficiency or allocative efficiency can be considered. *Technical efficiency* refers to the ability to obtain a maximum output that would be attainable when using a given amount of inputs, whereas *allocative efficiency* refers to the ability to use inputs in optimal proportions given their prices, which corresponds to maximizing the profit function (Coelli *et al.*, 2005). Efforts to enhance animal welfare may influence both a farm's allocative efficiency and its technical efficiency. A farm's access to specific inputs, such as pastoral land or labour to take care of the animals, may influence both its efficiency and farm animal welfare. Putting more effort in on animal welfare can sometimes reduce a farm's productivity and efficiency. However, such choices can

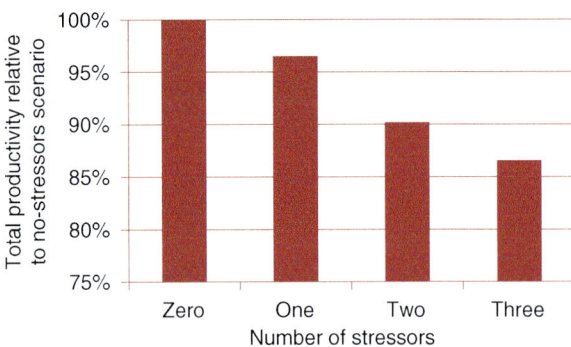

Fig. 5.3. An example illustrating that the presence of one, two or three animal stressors (high temperature, restricted space allowance or regrouping of animals) at the same time can influence total factor productivity of fattening pigs when compared to the no-stressors situation. (Author's calculation based on information provided by Hyun *et al.* (1998) and gross margin model presented by D'Eath *et al.* (2016).)

also be rational from the farmer's perspective because they can help him to meet objectives other than profit maximization. Hansson *et al.* (2018) illustrate that apparent inefficiency in dairy production may result from rational production decisions by farmers, based on ethical issues such as animal welfare concerns. Farmers may over-consume production inputs to achieve higher levels of animal welfare, because that may give them economic value not reflected in the regular output.

There is some evidence suggesting a positive association between livestock farm intensification and efficiency (e.g. Alvarez *et al.*, 2008; Cabrera *et al.*, 2010). In many cases, an opposite association is likely between animal welfare and intensification. Allendorf and Wettemann (2015) investigated process-based animal welfare indicators and the technical efficiency of German dairy farms. Their results indicated that, in particular, a higher percentage of cow losses, a higher replacement rate and a longer calving interval had, at their mean values, a negative marginal effect on the technical efficiency of the sample farms. By contrast, a lower age of first calving, a higher in-milk performance and a higher somatic cell count were positively correlated with technical efficiency. However, they found that efficient dairy farms did not always correspond with recommended values concerning animal welfare criteria and the effect was not always monotonic. Barnes *et al.* (2011) found that farms with low rates of lameness (<10% of the dairy herd) tended to have significantly higher technical efficiencies than those with lameness rates of above 10%. Low-lameness farms were inefficient in terms of labour and stocking density, but this was outweighed by the gain in milk yield obtained on these farms, thus supporting a whole-farm approach to farm efficiency analysis.

5.5 Empirical Evidence on the Cost of Compliance with Animal Welfare Standards

That higher welfare comes at economic cost is neither surprising nor unacceptable; one must expect that higher quality (hence higher value) goods cannot be gained without some additional cost. A central issue regarding productivity impacts of improvements on animal welfare standards can be examined by using McInerney (2004) model discussed above. The economic cost of adopting a higher animal welfare standard is usually a trade-off – the sacrifice in livestock productivity and simultaneous improvement in animal welfare and monetary and non-monetary benefits of this improvement. From the farmer's perspective, an economically critical factor is whether the benefits to the farm are larger than the costs.

Besides direct impacts on productivity and variable input usage, changes leading to prohibition of a production system and the way that animals are kept can be of vital importance. When elevating standards require changes to fixed inputs, farmers may face *sunk costs*. Sunk costs do not influence the optimal choice even if they may impact the farm's profit. For instance, abandoning sow-farrowing crates may require that housing is redesigned and modified.

While additional costs to housing may be reasonable, major elements of fixed capital may have to be scrapped or made subject to substantial reinvestments or modification costs before they have reached the end of their productive life. If this type of change is required at short notice, farmers may not be able to gain financial compensation for their scrapped inputs. Then farmers whose capital items are new can suffer higher losses from early scrapping than those whose capital items are near the end of their productive life.

Because the issues raised are often species-specific or farming system-specific, a formal assessment is required to judge whether the net economic effect, taking into account reduced productivity and higher value of livestock product from such a change, is positive or negative. A few studies have assessed farmers' cost of compliance with animal welfare standards. Menghi *et al.* (2011) assessed costs associated with meeting European Union legislation in the field of animal welfare, food safety and environmental protection. They included operational costs (e.g. variable input and labour costs), investments, forgone production and profit (opportunity cost), and private transaction costs, including administrative costs. They also looked at benefits for farmers, including savings in input use, disinvestments, investment support, additional revenues and subsidies, and extension programmes financed with public funds. The cost items included land, labour and capital as well as non-factor costs such as feed, contractors, maintenance and depreciation of machinery and buildings. In practical terms, they considered factors such as group housing of sows, increased minimum space allowance, renovation of floors, application of enrichment materials, increased usage of fibre in feed and construction of light openings in broiler buildings. Their assessed costs for animal welfare regulations plus several directives to combat animal diseases and foodborne pathogens, contributing partly to animal welfare, were less than 2% of the total production cost of dairy milk and beef in the EU Member States, up to 3% in sheep, up to 8% in pigs and up to 4% in broilers. Earlier, Grethe (2007) had argued that the cost of compliance with year 2007 standards was 6% (at maximum) of production cost in pigs, up to 20% in egg production and about 10% in broiler production.

More recently, Spiller *et al.* (2015, cited in Grethe, 2017) estimated the costs of elevating animal welfare standards in Germany in the following fields: access of all livestock to different climate zones (preferably including an outdoor climate or grazing area for dairy cattle); provision of different functional areas with varying floor coverings; provision of installations, substances and incentives for species-specific activities, feed intake and grooming activities; provision of sufficient space; and no tail docking in pigs and no beak trimming in poultry. To substantiate these guidelines, existing minimum requirements were defined for fattening pigs, bullocks, chickens and laying hens in Germany. Their estimates were 2–5% of the total production costs for cow's milk, 1–18% for eggs, 9–22% for chicken meat, and as much as 18–27% for beef and 28–41% for pork. Majewski *et al.* (2012) provide another empirical example of how to identify costs and benefits related to enhanced animal welfare standards.

Only a few studies provide empirical post-intervention evidence on whether animal welfare pays off to the farm, and the evidence is varied. Henningsen *et al.* (2018) found large variations in both gross margin and animal welfare indicators in Danish pig farms. The relationship between these two indicators was weak, but it tended to be slightly positive. A possible explanation was that management had a major influence on both economic performance and farm animal welfare so that good farm managers were able to comply with all animal welfare regulations while achieving a high economic performance.

Sampolahti (2013) found that the Welfare Quality scores and gross margins of Finnish pig farms tended to correlate positively. However, the statistical significance of correlations was quite weak. The most significant correlations with the gross margin were found for factors concerning feeding, adequate number of drinking nipples, prevention of injuries and diseases, and good relationship between the pigs and their caretaker. In particular, labour input was found to correlate with both the financial results of the farm and animal welfare. This suggests that labour may play an important role in both financial performance and animal welfare.

Gocsik *et al.* (2015) compared the economic feasibility of production systems with different levels of animal welfare in the broiler, laying hen and fattening pig sectors in The Netherlands. Their results suggest that the main determinant of economic feasibility in each sector is the producer price. Besides the level of the price premium obtained for animal welfare, the volatility of this premium is important when farmers decide whether to convert to an alternative system. They concluded that the broiler sector had the best perspective in the short to medium term for developing markets for animal welfare-oriented products. In the fattening pig sector, they concluded that higher price premiums or additional subsidies were needed to cover the conversion costs of production systems, and the laying hen sector had the worst prospects for improving animal welfare.

Gocsik *et al.* (2016) compared conventional, middle-market and top-market broiler production systems in The Netherlands. The major system attributes that differentiated between production systems were broiler type, stocking density and outdoor access. The Welfare Quality (Welfare Quality®, 2009) score of extensive outdoor (733) and organic systems (698) was below that of the middle-market systems. With regard to cost-efficiency, when shifting from conventional to an alternative system, middle-market systems (8.37) outperformed the extensive outdoor (3.90) and organic systems (1.03). They concluded that the middle-market systems could be attractive for farmers due to their high cost-efficiency and the flexibility to revert to the conventional system (Gocsik *et al.*, 2016).

More commonly, studies have elaborated specific animal welfare improvements or costs related to these. The financial implications of space allowance is one of the most frequently addressed issues. Bornett *et al.* (2003) found that when space allowance of fattening pigs was increased by 60%, the rearing costs per kilogram of pig meat were unchanged for the free-range pig-fattening system but rose by 5% for the pig-fattening system with fully-slatted floor, thus suggesting that it may be possible to improve pig welfare with reasonable additional costs. Herva *et al.* (2015) found that increasing space allowance for fattening bulls in Finland by keeping fewer animals in the same pen reduced fattening farm's profit. The potential of animal welfare improvements to generate additional revenue was therefore important when considering whether to reduce stocking density. De Vries *et al.* (2016) found that an optimal dairy barn stocking density in the USA varied depending on the combination of inputs, and it was very sensitive to changes in the size of the milk loss and milk and feed prices, which may arise because of changes in stocking density. They found that overstocking was profitable under plausible economic conditions in the USA and a trade-off will occur between economically optimal stocking density and animal welfare in some situations.

Heise *et al.* (2018) observed no significant effect of participation in an animal welfare programme on the *perceived* economic success of farmers. Odermatt *et al.* (2019), by contrast, analysed farms participating in a support programme which provided a compensation for group-housed dairy systems with a comfortable lying area separated from the feeding area or for cows receiving regular exercise outdoors in the winter and in pasture during summer. They found that the implementation of higher welfare standards can have a positive economic effect on the farm. Farms participating in the latter programme tended to reduce their veterinary costs by 2%, and farms participating in both programmes by 10%.

Various diseases can reduce animal welfare. While the financial impact of untreated diseases can vary, there are interventions to control diseases while simultaneously reducing financial losses. However, at least in poultry there appears to be sufficient scientific evidence on financially optimal disease-prevention and treatment patterns (e.g. Jones *et al.*, 2018), although there are gaps in knowledge for other species. Besides monetary factors, non-monetary factors such as benefits in terms of job satisfaction have been reported to motivate farmers to adopt interventions (Tremetsberger and Winckler, 2015). Impaired longevity of animals such as sows or dairy cows can arise from compromised welfare. The issue has been studied widely and studies indicate that premature replacement can result in high economic costs (e.g. Niemi *et al.*, 2017; Kerslake *et al.*, 2018). Kerslake *et al.* (2018) illustrate that problems attributed to non-pregnancy, mastitis, udder problems, calving trouble, and injury or accident can cause substantial economic losses to dairy farmers.

Pasture-based production systems have a positive impact on dairy cattle welfare, but maintaining pastures can be costly. French *et al.* (2015) argued that the primary economic focus of pasture-based systems should be to maximize the length of the grass-grazing season. Provided that body condition targets can be met, they expected minimal effect of wintering system on dairy cow productivity and the economic differences only in production costs.

To illustrate some of the effects in more detail, a simple case study on farrowing systems is presented.

Case Study: Switching from sow-farrowing crates to crate-free systems

Confinement of sows into crates during farrowing and gestation is one of the criticized practices in livestock farming. The issue is economically relevant and has been studied over recent years by various research groups. Vosough Ahmadi *et al*. (2011), for instance, modelled welfare components (inputs), including extra space, substrate and temperature, the economic performance of conventional crates, simple pens and designed pens, subject to both managerial and animal welfare constraints. Their results indicated that, when using piglet survival rate in a linear programming model based on data drawn from the literature and incorporating costs of extra inputs in the model, the crates provided the highest annual net margin and the designed pens and the pens were in second and third place, respectively. The designed pens were able to improve their annual net margin once alternative reference points, following expert-derived production functions, were used to adjust piglet survival rates in response to extra space, extra substrate and modified pen heating. The non-crate systems then provided higher welfare and higher net margin for sows and piglets than did crates, implying the possibility of a win–win situation.

Jääskeläinen *et al*. (2013) studied the cost and benefits of switching from farrowing crates to pens. Based on the literature, they developed several scenarios on how this welfare improvement could be implemented. Three of these scenarios are represented here. Stillbirth rate was assumed to be higher in crates than in pen-housing scenarios, whereas the opposite was assumed for piglet mortality between birth and weaning. Because a farrowing pen is more animal-friendly than a crate, additional animal welfare support was available for farrowing pen systems at the time of the analysis. Farrowing pen systems were assumed to be more labour-intensive, incur higher variable costs and require more capital investment. Changing from crates to pens in currently operating farms involves modifying housing by rebuilding the pens and/or interior design of the farrowing compartment or rebuilding the entire pig house. Therefore, the modification costs and costs of scrapping current pens before the end of their productive life and the additional capital costs can be substantial if farms have invested recently in a farrowing crate system and must scrap the old system. The scenarios included the case where modifications were implemented only after the end of the productive life of the current system. In each case, the farmer was assumed to have a fixed number of sows.

Table 5.1 illustrates the costs and benefits of switching from farrowing crates to a farrowing pen system. In each case, the numbers represent the difference between a farrowing pen system and farrowing crates. These scenarios represent different options of *how* to adopt an animal welfare-improving change in the farm. The level of animal welfare is assumed to be the same in scenarios Pen 1 and Pen 3 once the change has been adopted. However, scenario Pen 3 adopts the change later than other scenarios and therefore represents a lower overall level of animal welfare. Scenario Pen 2 differs from Pen 1 by having a higher level of piglet mortality.

As scenario Pen 1 illustrates, switching from farrowing crates to farrowing pens is expected to increase housing costs (pens must be modified), labour costs (it takes more effort to take care of pigs) and other variable costs such as feed costs. The farm is also expected to receive an additional support payment when committing to pens as a more animal-friendly production practice, but this is insufficient to cover increased costs. Scenario Pen 2 illustrates a case where the yield is decreased because of an animal welfare improvement. In both scenarios 1 and 2, the net effect is negative, i.e. the increase in costs

Continued

Continued.

is more than what can be obtained as higher revenues. Scenario Pen 3 illustrates a case where the pens are introduced after a transition phase, after the current housing is at the end of its economic life. The fixed costs are lower in Pen 3 than in Pen 1 because the costs caused by abandoning facilities that are still in use are reduced. Therefore, it is of economic importance to the farm how the change is implemented.

Table 5.1. Parameters used in a case study examining costs and benefits of switching from farrowing crates to farrowing pens in Finland.

Parameter	Crate system	Pen 1 Rebuild pens	Pen 2 Rebuild pens + elevated piglet mortality	Pen 3 Build a new system when the current system has reached the end of its life
Stillbirths	9%	8%	9%	8%
Piglet mortality, from birth until weaning	14%	15%	19%	15%
Additional support, €/sow space/year	8	19.5	19.5	19.5
Change in the lactation feed costs	0%	7%	7%	7%
Change in labour input usage	0%	17%	17%	17%
Change in other variable costs, €/sow	80	93	93	93
Investment in new housing, €/sow space	3222	3558	3558	3428
Current value of investment (% of new)	100	75	75	0

Source: Jääskeläinen *et al.* (2013)

Figure 5.4 illustrates another aspect that is of importance; namely: because the financial margin of the farm is obtained by subtracting the costs from the revenues, relatively small cost increases can have substantial effects on the gross margin of the farm, and even more dramatic effects on the net margin. In scenario pen 1, the additional revenue required to break even with the crate-system housing was €96/sow space/year. This was more than the estimated annual gross margin of the farrowing crate system. Hence the change would erode all net profit and about 15% of compensation to fixed capital and labour costs. How-

Continued

Continued.

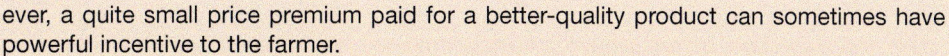

ever, a quite small price premium paid for a better-quality product can sometimes have powerful incentive to the farmer.

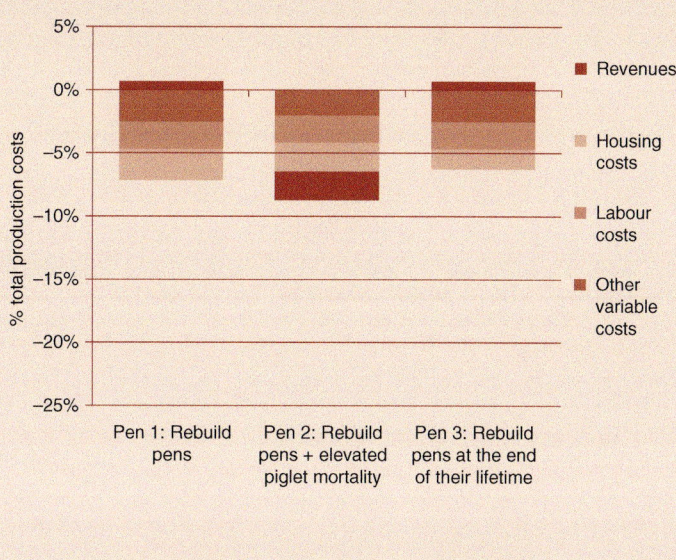

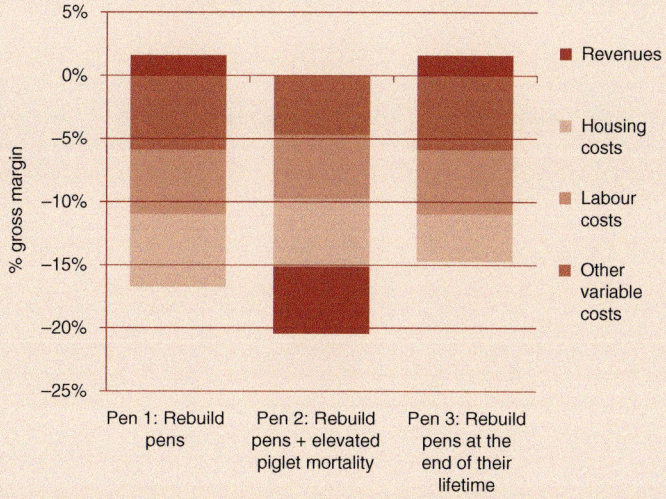

Fig. 5.4. Estimated impacts of switching from farrowing crates to farrowing pens as percentage of total production costs. (On the left, baseline €1487/sow space/year, and as the percentage of farm's gross margin on the right, €635/sow space/year.) (Jääskeläinen *et al.*, 2013)

5.6   Conclusions

In this chapter, we have examined how to analyse costs and benefits of animal welfare at the farm level. It is essential that farmers can gain monetary or non-monetary benefits (benefits to humans) when improving animal welfare because this can encourage them to adopt animal-friendly production practices. Some practices, such as those improving animal health, can provide benefits through improving productivity, whereas in other cases improving animal welfare typically involves additional costs because higher-quality goods cannot be gained without some additional cost. It is therefore important that the farmer who decides about complying with a higher animal welfare standard can gain additional revenue from the market, additional support or productivity improvements.

References

Allendorf, J.J. and Wettemann, P.J.C. (2015) Does animal welfare influence dairy farm efficiency? A two-stage approach. *Journal of Dairy Science* 98(11), 7730–7740. DOI: 10.3168/jds.2015-9390.

Alvarez, A., del Corral, J., Solís, D. and Pérez, J.A. (2008) Does intensification improve the economic efficiency of dairy farms? *Journal of Dairy Science* 91(9), 3693–3698. DOI: 10.3168/jds.2008-1123.

Barnes, A.P., Rutherford, K.M.D., Langford, F.M. and Haskell, M.J. (2011) The effect of lameness prevalence on technical efficiency at the dairy farm level: an adjusted data envelopment analysis approach. *Journal of Dairy Science* 94(11), 5449–5457. DOI: 10.3168/jds.2011-4262.

Bennett, R. (2012) Economic rationale for interventions to control livestock disease. *Eurochoices* 11, 5–10.

Bornett, H.L.I., Guy, J.H. and Cain, P.J. (2003) Impact of animal welfare on costs and viability of pig production in the UK. *Journal of Agricultural and Environmental Ethics* 16(2), 163–186. DOI: 10.1023/A:1022994131594.

Cabrera, V.E., Solís, D. and del Corral, J. (2010) Determinants of technical efficiency among dairy farms in Wisconsin. *Journal of Dairy Science* 93(1), 387–393. DOI: 10.3168/jds.2009-2307.

Coelli, T.J., Rao, D.S.P., O'Donnell, C.J. and Battese, G.E. (2005) *An Introduction to Efficiency and Productivity Analysis*, 2nd edn. Springer Science & Business Media, New York.

Dawkins, M.S. (2016) Animal welfare and efficient farming: is conflict inevitable? *Animal Production Science* 57(2), 201–208. DOI: 10.1071/AN15383.

D'Eath, R.B., Niemi, J.K., Vosough Ahmadi, B., Rutherford, K.M.D., Ison, S.H. *et al.* (2016) Why are most EU pigs tail docked? Economic and ethical analysis of four pig housing and management scenarios in the light of EU legislation and animal welfare outcomes. *Animal* 10(4), 687–699. DOI: 10.1017/S1751731115002098.

De Vries, A., Dechassa, H. and Hogeveen, H. (2016) Economic evaluation of stall stocking density of lactating dairy cows. *Journal of Dairy Science* 99(5), 3848–3857. DOI: 10.3168/jds.2015-10556.

French, P., Driscoll, K.O., Horan, B. and Shalloo, L. (2015) The economic, environmental and welfare implications of alternative systems of accommodating dairy cows during the winter months. *Animal Production Science* 55(7), 838–842. DOI: 10.1071/AN14895.

Gocsik, É., Brooshooft, S.D., de Jong, I.C. and Saatkamp, H.W. (2016) Cost-efficiency of animal welfare in broiler production systems: a pilot study using the Welfare Quality® assessment protocol. *Agricultural Systems* 146, 55–69. DOI: 10.1016/j.agsy.2016.04.001.

Gocsik, É.., Oude Lansink, A.G.J.M., Voermans, G. and Saatkamp, H.W. (2015) Economic feasibility of animal welfare improvements in Dutch intensive livestock production: a comparison between broiler, laying hen, and fattening pig sectors. *Livestock Science* 182, 38–53. DOI: 10.1016/j.livsci.2015.10.015.

Grethe, H. (2007) High animal welfare standards in the EU and international trade – How to prevent potential 'low animal welfare havens'? *Food Policy* 32(3), 315–333. DOI: 10.1016/j.foodpol.2006.06.001.

Grethe, H. (2017) The economics of farm animal welfare (October 2017). *Annual Review of Resource Economics* 9, 75–94.

Hansson, H., Manevska-Tasevska, G. and Asmild, M. (2018) Rationalising inefficiency in agricultural production – the case of Swedish dairy agriculture. *European Review of Agricultural Economics* 6, 1–24.

Hansson, H. and Lagerkvist, C.J. (2015) Identifying use and non-use values of animal welfare: evidence from Swedish dairy agriculture. *Food Policy* 50, 35–42. DOI: 10.1016/j.foodpol.2014.10.012.

Heise, H., Schwarze, S. and Theuvsen, L. (2018) Economic effects of participation in animal welfare programmes: does it pay off for farmers? *Animal Welfare* 27(2), 167–179. DOI: 10.7120/09627286.27.2.167.

Henningsen, A., Czekaj, T.G., Forkman, B., Lund, M. and Nielsen, A.S. (2018) The relationship between animal welfare and economic performance at farm level: a quantitative study of Danish pig producers. *Journal of Agricultural Economics* 69(1), 142–162.

Herva, T.J., Niemi, J.K., Peltoniemi, O.A. and Saatkamp, H.W. (2015) Welfare of beef cattle and farm net income: the complex relationship among housing, space allowance and subsidies. In: Herva, T. (ed.) *Animal Welfare and Economics in Beef Production*. Dissertation. Faculty of Veterinary Medicine, University of Helsinki, Helsinki.

Hyun, Y., Ellis, M. and Johnson, R.W. (1998) Effects of feeder type, space allowance, and mixing on the growth performance and feed intake pattern of growing pigs. *Journal of Animal Science* 76(11), 2771–2778.

Jääskeläinen, T., Niemi, J. and Karhula, T. (2013) *Häkkiporsituksesta luopumisen tuotannolliset ja taloudelliset vaikutukset*. Animal welfare centre, University of Helsinki, Helsinki, p. 54.

Jones, P.J., Niemi, J., Christensen, J.-P., Tranter, R.B. and Bennett, R.M. (2018) A review of the financial impact of production diseases in poultry production systems. *Animal Production Science*.

Kerslake, J.I., Amer, P.R., O'Neill, P.L., Wong, S.L., Roche, J.R. *et al.* (2018) Economic costs of recorded reasons for cow mortality and culling in a pasture-based dairy industry. *Journal of Dairy Science* 101(2), 1795–1803. DOI: 10.3168/jds.2017-13124.

Lusk, J.L. and Norwood, F.B. (2011) Animal welfare economics. *Applied Economic Perspectives and Policy* 33(4), 463–483. DOI: 10.1093/aepp/ppr036.

Majewski, E., Hamulczuk, M., Malak-Rawlikowska, A., Gębska, M. and Harvey, D. (2012) Cost-effectiveness assessment of improving animal welfare standards in the European agriculture. In: *Selected Paper prepared for presentation at the International Association of Agricultural Economists (IAAE) Triennial Conference*, 18-24 August 2012. Foz do Iguaçu, Brazil. Available at: http://purl.umn.edu/126741

McFadden, D. and Fuss, M. (1978) *Production Economics: A Dual Approach to Theory & Applications, Vol. 1: The Theory of Production*. Available at: https://eml.berkeley.edu/~mcfadden/ (accessed 17 June 2020).

McInerney, J.P. (2004) *Animal welfare, economics and policy*. Report to Defra, UK. Available at: https://webarchive.nationalarchives.gov.uk/20110318142209/http://www.defra.gov.uk/evidence/economics/foodfarm/reports/documents/animalwelfare.pdf (accessed 17 June 2020).

Menghi, A., de Roest, K., Porcelluzzi, A., Deblitz, C., von Davier, Z. *et al.* (2011) *Assessing farmers' cost of compliance with EU legislation in the fields of environment, animal welfare and food safety*. AGRI-2011-EVAL-08. Final report commissioned by the European Commission Directorate-General for Agriculture and Rural Development. Available at: https://ec.europa.eu/agriculture/sites/agriculture/files/external-studies/2014/farmer-costs/full-text_en.pdf

Niemi, J.K., Bergman, P., Ovaska, S., Sevón-Aimonen, M.-L. and Heinonen, M. (2017) Modeling the costs of postpartum dysgalactia syndrome and locomotory disorders on sow productivity and replacement. *Frontiers in Veterinary Science* 4(181), 1–12. DOI: 10.3389/fvets.2017.00181.

Niemi, J.K., Stygar, A., Jones, P., Clark, B., Frewer, L. *et al.* (2018) The economic costs of production diseases in pigs. Prohealth industry workshop. Ghent, Belgium. Available at: http://www.fp7-prohealth.eu/documents/98/PROHEALTH_27Nov2018_Pigs-08_Niemi.pdf (accessed 16 April 2019).

Odermatt, B., Keil, N. and Lips, M. (2019) Animal welfare payments and veterinary and insemination costs for dairy cows. *Agriculture* 9(1), 3. DOI: 10.3390/agriculture9010003.

Rushton, J. (2009) *Economics of Animal Health and Production: A Practical and Theoretical Guide.* CAB International, Wallingford, UK.

Sampolahti, T.H. (2013) Lihasikojen hyvinvoinnin yhteys tuotannon taloudelliseen tulokseen. Master thesis. Department of Economics and Management, Faculty of Agriculture and Forestry, University of Helsinki. p. 68. Available at: https://docplayer.fi/45992101-Lihasikojen-hyvinvoinnin-yhteys-tuotannon-taloudelliseen-tulokseen.html (accessed 17 June 2020).

Spiller, A., Gauly, M., Balmann, A., Bauhus, J., Birner, R. *et al.* (2015) Wege zu einer gesellschaftlich akzeptierten Nutztierhaltung. *Ber. Landwirtsch. Sonderh* 221, 1–172.

Tremetsberger, L. and Winckler, C. (2015) Effectiveness of animal health and welfare planning in dairy herds: a review. *Animal Welfare* 24(1), 55–67. DOI: 10.7120/09627286.24.1.055.

Vosough Ahmadi, B., Stott, A.W., Baxter, E.M., Lawrence, A.B. and Edwards, S.A. (2011) Animal welfare and economic optimisation of farrowing systems. *Animal Welfare* 20, 57–67.

Wallenius, E., Kauppinen, T., Raussi, S., Heinola, K. and Niemi, J. (2020) Eläinten hyvinvointia edistävät toimet hyvinvointimerkin takana. *Suomen Maataloustieteellisen Seuran Tiedote* (38), 1–12.

Welfare Quality® (2009) Welfare Quality® assessment protocol for poultry (broilers, laying hens). Welfare Quality® Consortium, Lelystad, The Netherlands. Available at: http://www.welfarequality.net/media/1019/poultry_protocol.pdf (accessed 17 June 2020).

6 Poultry Breeding for Sustainability and Welfare

ANNE-MARIE NEETESON,* SANTIAGO AVENDAÑO, AND ALFONS KOERHUIS

Aviagen Group

Summary

Animal breeding for welfare and sustainability requires improving and optimizing environmental impact, productivity, robustness and welfare. Breeding is a long-term exercise at the start of the food chain with permanent cumulative outcomes, disseminated widely. This chapter explains, with a focus on poultry, breeding programme design and how broadening breeding goals and managing trait antagonism results in balanced breeding and more robust animal populations. Breeding progress in skeleton and skin health, physiology and body composition, and behaviour are addressed. The economic impact of welfare and environmental improvements is worked out, and the ethical and societal aspects of genetic improvement are put into perspective. The consideration of feedbacks of all stakeholders, including customers and the wider society, is crucial. For each crossbreed, breeders will continue to improve overall welfare, health, productivity and environmental impact, but between the crossbreeds there will be clear differences answering specific demands of concepts and brands.

6.1 Introduction

Animal breeding and genetics are situated at the start of the food supply chain, with a gene flow originating from the breeding programme and disseminating permanent and cumulative genetic change through the production base. This chapter addresses the global framework of which animal breeding, economics and welfare are part, and the technical and organizational details of

*Corresponding author: aneeteson@aviagen.com

breeding for sustainable poultry meat production. It will focus on poultry, using examples from broiler and turkey breeding.

6.1.1   Relevant global developments

The world's human population will grow to over 9 billion by 2050, with long-term increases in meat demand, limited availability of natural resources such as agricultural land and water, and concerns about environmental impact of livestock production (Foresight, 2011; Herrero and Thornton, 2013; Herrero *et al.*, 2013; LEAP (Livestock Environmental Assessment and Performance Partnership, 2015; United Nations, 2015; Bryden, 2016).

At the same time, with societies becoming more affluent, alternative production systems with various schemes targeting thresholds for growth rates lower than 50 g/day and/or specific requirements regarding welfare attributes, coloured feathering of birds and/or ranging behaviour are gaining momentum (Neeteson and Avendaño, 2016b; Avendaño *et al.*, 2017). These labelling schemes require alternative production systems or new suitable crossbreeds (e.g. Kip van Morgen (CBL, 2013) and Beter Leven (beterleven.dierenbescherming.nl) in The Netherlands, Für mehr Tierschutz (tierschutzlabel.info) in Germany, and RSPCA Assured (rspcaassured.org.uk) in the UK. In addition to these consumer requirements, consumer choice, responsible use of antibiotics, dietary health and sustainability are drivers of food purchases in urban and prosperous societies.

6.1.2   Sustainability

Economy, welfare and breeding are an integral part of sustainability. The official definition of sustainability is 'meeting the needs of the present without compromising the ability of future generations to meet their own needs' (Brundtland report, 1987). Sustainability is mostly referred to via the three Sustainability Pillars (United Nations General Assembly, 2005): environment, society and economy. Later, 'society' has also been referred to as 'ethics'. Animal welfare is an integral part of the ethics/society pillar; economy occupies a pillar of its own. Breeding contributes to sustainability by delivering genetic potential to match both environmental (through lower resource utilization and carbon footprint), economic (through increased profitability of the wider industry) and welfare requirements (through meeting societal requirements).

6.1.2.1 Sustainable breeding

The framework of sustainability in animal breeding was coined in 2003. Motivated by the apparent distance between animal production in its wider sense and consumers, breeders, scientists, welfare organizations, ethicists, sociologists and economists developed the concept of sustainability in animal breeding in EC-funded efforts (Neeteson-van Nieuwenhoven *et al.*,

2006). As a first step, the ethical, legal and societal implications of farm animal breeding were described (Neeteson *et al.*, 1999). Breeders were guided by ethicist Professor Peter Sandøe, who proposed that they must investigate and decide where animal breeding can make a difference. In the second phase, socio-economic studies explored cultural differences, animal welfare, ethics and global economy in relation to animal breeding. At the same time, socio-economic specialists provided a critical forum for the specialists in ruminant, pig, poultry and aquaculture breeding who independently identified similar areas across species where animal breeding can play a role: welfare and health, food safety, genetic diversity, efficiency, environment and product quality. The end results were a definition of and scenarios for sustainable farm animal breeding and reproduction (Liinamo and Neeteson-van Nieuwenhoven, 2003; FABRE TP (Sustainable Farm Animal Breeding and Reproduction Technology Platform), 2008). In a third step, this was formalized in a Code of Good Practice for Sustainable Farm Animal Breeding and Reproduction (Code-EFABAR; responsiblebreeding.eu; EFFAB, 2018a, EFFAB (European Forum of Farm Animal Breeders), 2018b), which is updated every 3 years and adopted by many breeding organizations. The Code is a pledge by the breeding companies with which they show how 'selection for genetic potential in production is balanced by appropriate attention to reproduction, health and welfare-related traits' (Besbes and Neeteson-van Nieuwenhoven, 2006). Poultry breeding companies now take this code a step further by adding an auditing component with Code-EFABAR Poultry Assured (CEPA; EFFAB (European Forum of Farm Animal Breeders), 2018c).

6.1.2.2 Sustainable animal production

Animal production sectors have articulated their sustainability actions for their respective species through several forums. In beef, there is an active Global Round Table for Sustainable Beef (grsbeef.org). The International Dairy Federation and the UN Food and Agriculture Organization (FAO) have launched the Sustainable Dairy Development Goals (Bellamy and Bogdan, 2016). The International Poultry Council has adopted the Poultry Sustainability Pillars and has linked these to the UN Sustainable Development Goals (SDGs) (Declaration of São Paulo; IPC, 2017, IPC (International Poultry Council), 2019). The International Egg Commission announces it will work out how the egg sector contributes to the UN SDGs (IEC (International Egg Commission), 2018). Recently, stakeholder platforms like the International Poultry Welfare Alliance (IPWA) and US Round Table for Sustainable Poultry and Eggs (US-RSPE) were launched (USPEA, 2018).

6.1.2.3 Breeding for animal welfare

For a poultry breeder, welfare can be defined as developing suitable birds for a range of production systems via balanced breeding from large gene pools, handling birds with care and respect, ensuring health, welfare and productivity. At the same time it should support producers with the latest management and technical advice, and demonstrate corporate responsibility while meeting global food demand. This principle is described as the commitments to (i) health, food safety and food security by ensuring a safe and secure supply of healthy birds to help feed a growing global population; (ii) biodiversity by providing a sustainable supply of breeding stock from pedigree programmes with large diverse gene pools; (iii) balanced breeding programmes in which simultaneous selection takes place for a large number of production, health and welfare traits; (iv) tailored management and stockmanship via extensive management support as management factors contribute significantly to the health, welfare and performance of a flock; and (v) transparency, communication and engagement: engaging with stakeholders through active, transparent communication and close cooperation with customers, associations and wider society (Aviagen Group, 2017).

6.2   Animal Breeding

6.2.1   The domestication and breeding of animals

6.2.1.1 Domestication

When hunter-gatherer populations became farmers, domestication of farm animals started, for example domesticated chickens originate from the red jungle fowl (*Gallus gallus*, 2500–2100 BC; Laughlin, 2007). Domestication gradually developed into animal breeding, which is the selective mating of animals with desirable genetic traits, to maintain or enhance these traits in future generations (FABRE TP (Sustainable Farm Animal Breeding and Reproduction Technology Platform), 2008; Nature, 2018).

6.2.1.2 The gene flow structure

Figure 6.1 shows two entangled pyramids representing the broiler gene flow structure. The bottom pyramid contains requirements to develop and maintain a breeding programme – a gene pool including pure lines contributing to commercial broilers or turkeys and non-commercial lines (e.g. experimental and control lines), and the considerable support systems of modern genetics. Poultry breeders keep a wide range of pedigree lines (for instance, in Aviagen >30 in broilers, >40 in turkeys; Defra (Department for Environment, Food and Rural Affairs), 2010). The inverted pyramid shows how genetic improvements made in poultry pedigree lines are disseminated to the rest of the chain through a series of multiplying generations into a variety of crossbreeds at parent and broiler/commercial turkey level to meet the needs of different markets. A crossbreed is typically made up of four different types of pedigree lines (Laughlin, 2007; Aviagen Group, 2016).

Animal breeding is the optimization of internal and external factors: maximizing the breeding objective(s) taking account of the opportunities and challenges from the outside world (economy, environment, society), and the intrinsic opportunities and limitations from animal populations, structures and opportunities to correctly measure and weigh the genetic component of

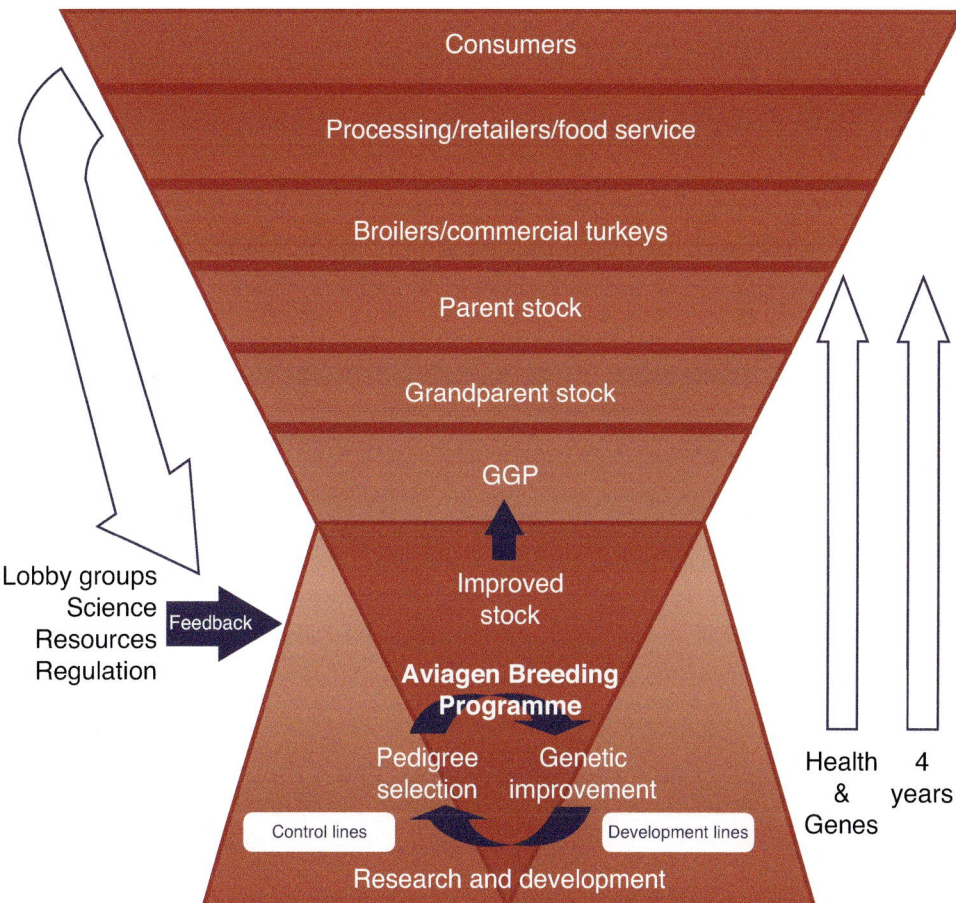

Fig. 6.1. Diversity of breeding stock, multiplication pyramid and feedback mechanism of Aviagen breeding programmes. GGP=great grandparent stock. (From Laughlin, 2007)

the desired characteristics. Successful breeding requires accurate prediction of breeding values with data recording, novel trait development, optimal breeding programme structure and design, the use of latest knowledge in analytics applied to genetics and genomics, and the ability to manage huge datasets at high throughput. In addition, the size of the breeding populations matters as genetic progress is a direct function of selection intensity.

Animal breeding is a long-term exercise. Although economic evaluation is useful to indicate the broad direction, the breeding programme must have a broad and long-term focus. McKay *et al.* (2000) predicted that future selection progress also will include closer monitoring of physiological systems to ensure an integrated response to multiple traits. Neeteson-van Nieuwenhoven *et al.* (2013) queried animal breeding experts worldwide about the relative importance of trait groups in breeding goals in 1950, 1990, 2015 and 2025, showing that the complexity of breeding goals increases over time: the relative focus on productivity decreases and objectives such as efficiency (productive and environmental) and animal robustness increase. Breeding decisions require a long-term view, which at least should include a 10-year perspective – the consideration of feedback of all the industry stakeholders on a continual basis is crucial (EPB/EFFAB (European Poultry Breeders/European Forum of Farm Animal Breeders), 2009). Figure 6.1 shows the feedback from customers and the wider society via an arrow to the left of the pyramids.

An important aspect of animal multiplication is healthy propagation of breeding stock. In cattle and pigs, artificial insemination was introduced to eliminate venereal diseases. In most animal breeding programmes the biosecurity level is very high. Biosecurity is a condition *sine qua non* in global poultry breeding. Any breeding programme must commit to deliver genetic stock free of transmissible diseases including *Mycoplasmas*, *Salmonellas* and Avian Influenza (AI). Strict biosecurity procedures must be in place for disease prevention and thousands of serology and bacteriology samples are analysed annually to ensure freedom from *Salmonella* and *Mycoplasma*. Avian Influenza represents the single most significant disease challenge, risk and threat to the poultry industry globally, which in recent years had serious consequences to poultry trade and the poultry industry leading to the destruction of millions of domestic birds across the world (OIE (World Organisation for Animal Health), 2017). The arrows on the right illustrate the biosecurity of the pyramid and its timeframe of 4 years from pedigree to poultry meat production.

6.2.2   Breeding programme design

The optimal design of the breeding programme is crucial for its sustainability. Maximizing genetic gain while controlling inbreeding rates is an approach that ensures long-term sustainability of the breeding programme. A sustainable breeding programme requires continuous improvement of selection accuracy, novel recording technology, and understanding the genetic and, where possible, physiological relationships between traits of interest across the variety of lines in the gene pool (e.g. both conventional and slow grow genotypes), as any product arising from these has to be both competitive and sustainable (EPB/EFFAB (European Poultry Breeders/European Forum of Farm Animal Breeders), 2009; Hill *et al.*, 2016; Avendaño *et al.*, 2017).

6.2.2.1 Animal identification

Genetic improvement is highly dependent upon sustained accurate data collection, which is paramount for reliable predictions of genetic merit. Data are gathered from pedigree animals and their relatives (ancestors, siblings), including information on traits which cannot be measured on the animal itself, e.g. egg production of a cockerel. In poultry breeding, a full pedigree structure of populations dates back to the late 1970s. Information recorded on the selection candidate and its relatives are included to predict genetic merit for all the traits in the breeding goal. In poultry breeding, the recording of health and welfare traits dates back to before the 1970s (Aviagen Group, 2016).

6.2.2.2 Trait development

In a multi-trait breeding goal, there is a wide range of traits with different levels of complexity. There are some traits that are relatively easy to measure and they represent the final outcome of a biological process; for example, live weights, physical examination of broilers and turkeys for leg and skeletal defects, egg production or fertility. On the other hand, there are traits that require significant technological developments because they underlie complex biological processes; for example, feed efficiency and behaviour, immune response and gut function. Transponder technology has allowed the recording of feed intake and feeding behaviour in large groups (e.g. Howie *et al.*, 2011, in chickens). X-ray technology has contributed to the improvement of leg health in broilers through the detection of the clinical and subclinical incidence of tibial dyschondroplasia (TD). The recent introduction of 3D imaging technology in broiler breeding is going to provide unprecedented opportunities to manage trade-offs between traits and will benefit, in particular, welfare, processing and meat quality traits and balance of the breeding programme in general (Howie *et al.*, 2011; Kapell *et al.*, 2012b; Avendaño *et al.*, 2017). Clearly, as more traits related to biological functions are included in the breeding goal, there will be a need to develop novel ways to measure underlying traits in the selection candidate through live indicators (e.g. biomarkers) or imaging technology.

6.2.2.3 Quantitative genetics and traditional genetic evaluation

Traditionally, genetic evaluation is based on phenotypic records, which incorporate genetic information through the family relationship and estimates of the relationships (genetic variances, co-variances) among the traits. The information on the animal, its parents, sibs and earlier generations are each combined optimally, taking account of all these aspects including each trait's heritability. With specifically designed models and analyses (e.g. Best Linear Unbiased Prediction (BLUP)), this wealth of information is used to predict breeding values (Hill *et al.*, 2016).

6.2.2.4 Genomics

Hill *et al.* (2016) explain that, when assessing individuals without their own or progeny records (in particular unproven males for sex-limited traits and for post-mortem recorded traits), information from the ancestors does not include the effect of which unique sample of DNA an individual gets from the father or from the mother (Mendelian sampling). Genomics information allows prediction of the Mendelian sampling term in the absence of an individual animal's own phenotypic record, hence allowing the differentiation of candidates from the same male and female parent, which otherwise would have the same breeding value. Simultaneous estimation of marker-associated effects across the whole genome creates genome-wide breeding value estimates, an approach developed by Meuwissen *et al.* (2001) ('genomic selection'). Genomic selection is implemented in many dairy breeding programmes and is routine in pig and poultry (e.g. Forni *et al.*, 2011; Avendaño *et al.*, 2010, 2012). Genomic selection combines pedigree, molecular and phenotypic data within the classical BLUP structure to maximize the accuracy of choosing selection candidates, allowing comparison of predicted breeding values of each individual directly, despite being based on very different amounts and types of data; as with BLUP, based on production data alone. Adding genomic prediction to traditional quantitative genetics evaluation in poultry increases the selection accuracy of 10–50% depending on the trait. The introduction of genomics into breeding programmes has thus increased the possibilities to select optimally balanced animals and the accuracy of selection for all traits, including those related to welfare (Aviagen Group, 2016; Hill *et al.*, 2016).

6.2.3   Breeding goals

A breeding goal defines the desired change in an animal population in terms of the traits of any kind of interest and their relative importance (Neeteson-van Nieuwenhoven *et al.*, 2013).

6.2.3.1 Broadening breeding goals

Breeding goals must be long term to achieve steady improvements over time, focus and improve the individual components, and weigh the optimum. Breeding goals also have to be adapted regularly, because the relative value of traits changes over time when there are changes in production conditions, market requirements and societal developments.

Better insights into the biological background of traits and of technology development lead to novel options for selection so that improving novel traits becomes feasible. Broiler breeding goals have expanded vastly in the last three decades, combining productivity and biological efficiency with liveability, robustness, adaptability and reproduction (Neeteson-van Nieuwenhoven *et al.*, 2013; Aviagen Group, 2016). Welfare-related traits represent 18–33% of the relative weight in the breeding goal in broiler breeding (Hiemstra and Ten Napel, 2013). This includes traits relating to skeletal integrity (e.g. leg strength, walking ability, TD), contact dermatitis, heart and lung function and liveability. In future, the process of breeding goal expansion is not expected to stop but to continue even further.

6.2.3.2 Sustainability of breeding goals

Whether livestock breeding goals are sustainable into the future depends on the ability of breeding organizations to balance the increasingly complex demands for overall sustainability: food security and economic survival; environmental load; population diversity; food safety; and health, welfare and ethical considerations. As explained, the further development of technology makes it increasingly feasible to aim for genetic improvement of productivity, product quality and animal adaptability simultaneously.

Hill *et al.* (2016) illustrate the principle of multi-trait genetic selection in the presence of trait antagonism in meat chicken breeding and demonstrate that when traits are included in a broad breeding goal and balanced selection is applied, the desired direction in each trait can be and has been achieved. In the next section, we explain the major decisive intrinsic components of a balanced breeding strategy: sufficient and sustainable management of antagonistic traits and of genotype by environment interaction to improve robustness.

6.2.4   Trait antagonism

In complex breeding programmes, including traits related to production, health, reproductive fitness, welfare, robustness and adaptability, there will often be traits that are antagonistically related. A classic example often quoted is that improvements in production traits lead to undesirable health-related consequences (e.g. Rauw *et al.*, 1998; Dawkins and Layton, 2012; Hocking, 2014). In the past, in broilers, increases in incidences of ascites and skeletal abnormalities were reported 'in birds intensively selected for growth rate' (Chambers, 1990). The way to handle these antagonisms is through estimating the genetic correlations between trait groups in the breeding goal (e.g. between biological performance and welfare or product quality) allowing improvements of antagonistically related traits simultaneously.

6.2.4.1 Antagonism in genetic terms

Avendaño *et al.* (2017) explained that, from a breeding point of view, the key parameter controlling the extent of the genetic antagonism between traits in a breeding goal is their genetic correlation (GC): '[T]he GC measures the extent to which two traits are controlled by the same genes. A favourable GC means that the genes controlling both traits have the same effect on each trait while an unfavourable or antagonistic correlation means that the effect is the opposite in each trait.'

6.2.4.2 Managing trait antagonisms

With an example from their breeding programme, Avendaño *et al.* (2017) illustrate the genetic antagonism between feed conversion rate (FCR – the rate at which feed is converted into bird live weight) and the percentage of eggs which successfully hatch to produce a viable chick (hatch %). Figure 6.2 shows the estimated breeding values (EBVs) for 1385 birds from a pedigree batch as deviations from the population mean. In this dataset, the GC between hatch percentage and FCR is 0.27, typical for the antagonism between broiler and breeder performance traits. The dotted double-arrowed trend line indicates that as we move further to the left we will find birds with better FCR (i.e. lower) but worse hatch percentage, while to the right we find birds with improved hatch percentage but poorer FCR (i.e. higher). In this example, FCR would get 0.012 (i.e. 12 g of feed per kg of live weight) worse per percentage improvement of hatch.

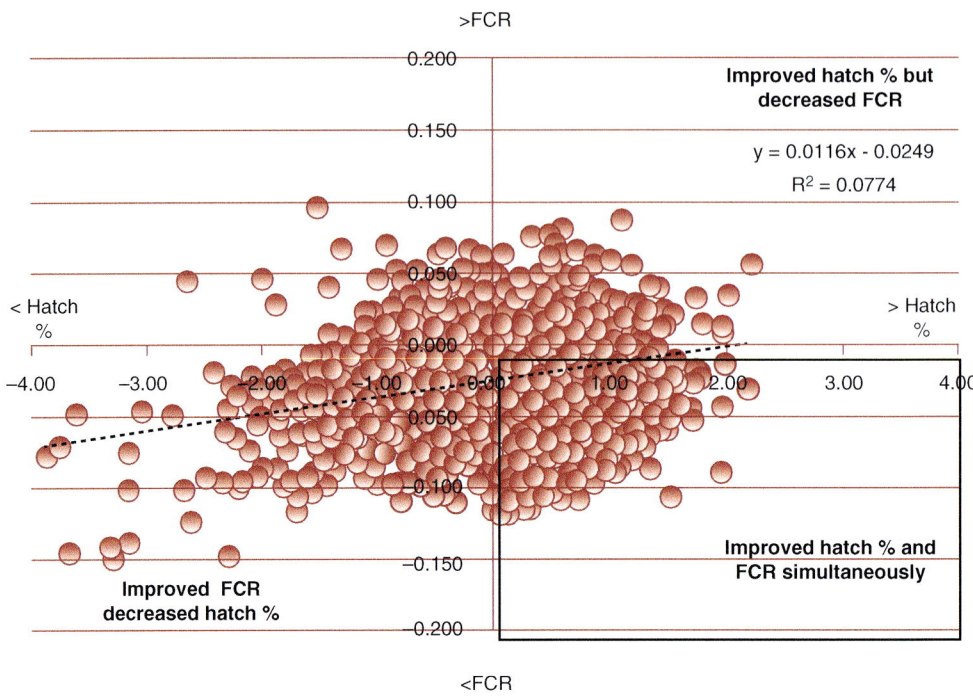

Fig. 6.2. Estimated breeding values (EBVs) for FCR (Feed Conversion Rate) (vertical axis) and hatch % (horizontal axis) as deviations from the population mean. (From Avendaño *et al*., 2017)

The solution to deal with this dilemma is to include both traits in the breeding goal and to select for birds which are good for both traits at the same time. The birds with breeding values in the right bottom box are the birds of this pedigree batch with better EBVs than the population average for both traits at the same time.

The current improved ways of recording both performance traits (e.g. body weight, breast yield %, FCR) and welfare-related traits (e.g. leg health and contact dermatitis) allow us inclusion of all these traits in the breeding goal, and to improve them simultaneously in the long term (management of antagonisms or trade-offs in animal breeding). This is illustrated by, for example, Kapell *et al*. (2012a, Kapell *et al*., 2012b) and confirms that 'continued genetic improvement for production efficiency and minimization of demand on resources is feasible without sacrificing animal health and welfare in a multi-trait selection set up including both fitness and production traits' (Hill, 2016).

6.2.5   Robustness

Current and future animal populations will have to perform with increasing efficiency in a wide range of environments, characterized by varying management practices, feed quality (form and density) and gut and immune challenges. In these environments, substantial genotype × environment (G×E) interactions will exist (Neeteson-van Nieuwenhoven *et al.*, 2013; Hill *et al.*, 2016).

6.2.5.1 The relevance of robustness

The robustness component of genetic potential is particularly relevant when meat production expands globally in a wide range of production regimes (OECD/FAO, 2018). The pedigree populations are normally kept at high standards of management, health and biosecurity. Thus it is crucial to include elements of differential production to the pedigree programme to minimize the G × E interaction and ensure the expression of the genetic potential in many varying local environments.

6.2.5.2 Breeding for robustness

One approach is to base selection at the pedigree level on records gathered in a commercial production environment. Pedigree populations can best express their genetic potential for productivity in high-input conditions. As highlighted before, animals raised in biosecure conditions are important to fulfil the other important role of breeding: dissemination of animals with a high health status. Methods used are combined crossbred and purebred selection (CCPS) (Wei and Van der Steen, 1991) and commercial sibling testing (CST) (Kapell *et al.*, 2012a). With CST, selecting on data from contrasting (pedigree and commercial) environments improves robustness. Sibs of selection candidates are grown in non-biosecure commercial conditions, assessing gut health, digestive and immune function along with liveability, growth and uniformity. The result is improved productivity in both environments, increased animal adaptability to the wider range of management circumstances they may encounter in the field, and more robust populations with higher liveability and uniformity.

6.3 Animal Welfare and Breeding

For reasons of practicality, we will first concentrate on welfare aspects where *genetics* is the major component: balanced breeding programmes. In a later section (6.6) we will also address other societal and ethical aspects influencing animal breeding.

6.3.1 Welfare improvements from breeding and management

For decades, breeding goals have been broad and included traits related to welfare. Apart from the overall balance in a broadly defined breeding goal, breeding programmes include individual traits related to welfare: health-related traits like leg, foot, gut health, stress susceptibility, muscle metabolism;

liveability in general; and adaptability traits such as susceptibility to behavioural deprivation (stereotypies, lethargy), dominance, aggression, feather pecking, tail-biting and piglet savaging (after Rodenburg and Turner, 2012). A report by FAWC (2012) confirmed that, indeed, farm animal breeding programmes have met many of the concerns raised in an earlier FAWC (Farm Animal Welfare Council) (2004) report. As described in Sections 6.2.4 and 6.2.5, the knowledge and experience of genetic parameters in modern breeding companies is such that these conditions can be, and have been, met.

6.3.2 Breeding for antagonistic welfare and production traits

Avendaño *et al.* (2017) describe how the concept illustrated in Fig. 6.2 is applied to manage the genetic antagonism between biological performance and welfare-related traits. Figure 6.3 shows the range of GC (across a number of pedigree broiler breeding lines) between body weight (BWT) and breast yield (BY, %) and a range of welfare-related traits: leg-bone deformities (LD, %), gait score (GS), tibial dyschondroplasia (TD, %), footpad dermatitis (FPD, %), crooked toes (CT, %), liveability (LIV, %) and oxygen saturation level in blood (OXI, %). Clearly, all the correlations are, in absolute terms, below 0.35,

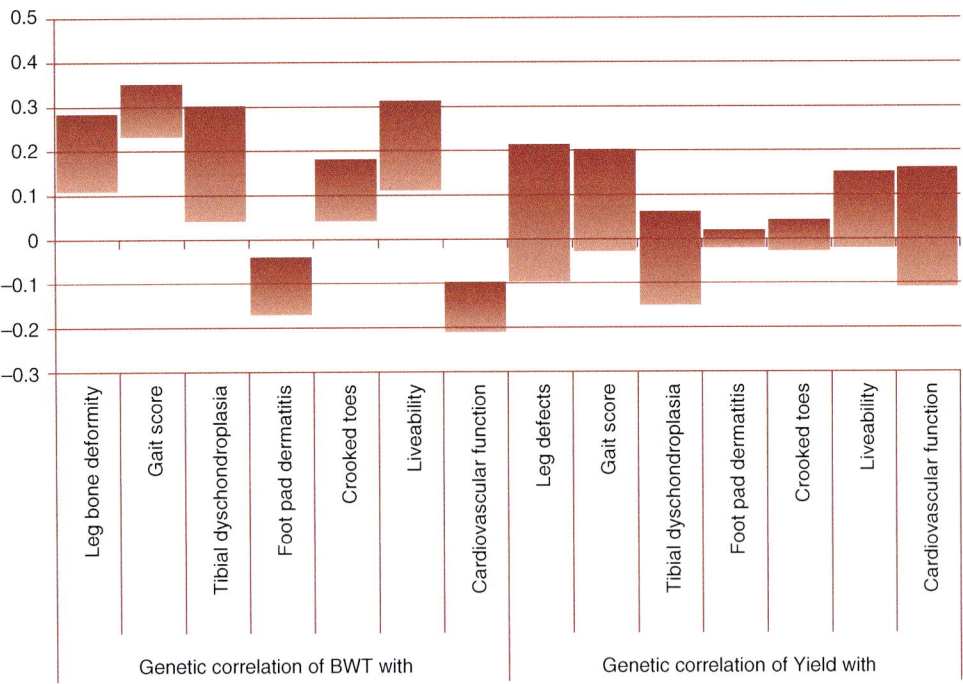

Fig. 6.3. Ranges of genetic correlations between live weight (BWT) and breast yield (%) with leg-bone deformities (%), gait score, tibial dyschondroplasia (%), footpad dermatitis (%), crooked toes (%), liveability (%) and oxygen saturation levels in blood (%; cardiovascular function). (From Avendaño *et al.*, 2017)

indicating that the extent of the antagonism is not extreme. In some cases the antagonism is insignificant (bars close to 0) or there is no genetic antagonism at all, e.g. between BWT and FPD, BY% and TD, FPD, CT and, in some cases, with OXI.

These GC ranges are essential – it means there are ample opportunities to improve biological performance and welfare-related traits simultaneously, even in the presence of genetic antagonisms, when both groups of traits are included in a broad and balanced breeding goal (Avendaño *et al.*, 2017). However, it also shows that the extent to which selection can be effective depends on whether: (i) records can be taken in a timely manner and, if these are proxies for the trait in question, are sufficiently closely correlated to it; (ii) there is significant genetic variation in that trait to enable its improvement; (iii) the trait is not so unfavourably correlated with other production, health and welfare traits that selection on it would be ineffective in improving overall productivity, efficiency and welfare; and (iv) the breeding programme can be appropriately structured to enable such selection to take place.

6.3.3   Review of poultry breeding for animal welfare

We will show the achievements in broiler breeding for welfare traits using some illustrative examples. Figure 6.4 (Avendaño *et al.*, 2017) illustrates that growth rate and leg strength have been improved simultaneously over the long term while the antagonistic genetic relationship between both traits

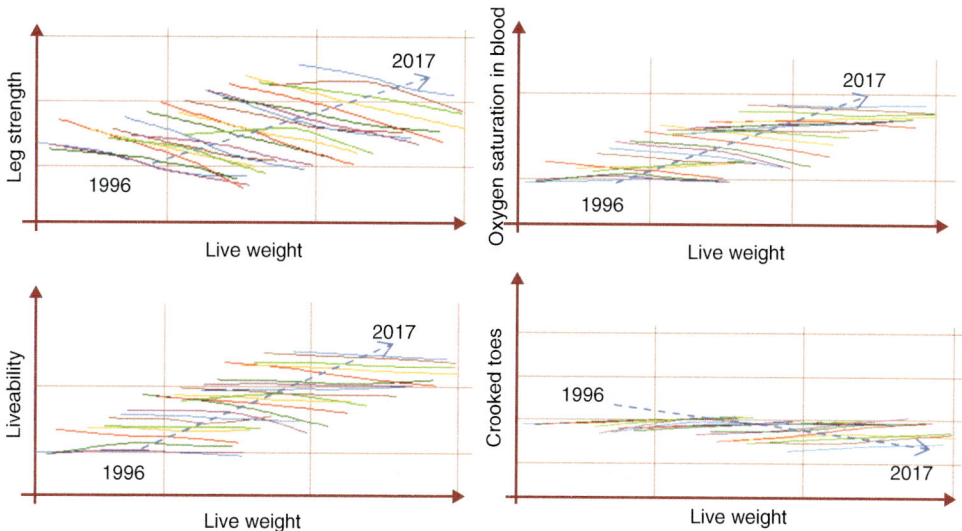

Fig. 6.4. Long-term relationships between live body weight and leg strength (%), liveability (%), oxygen saturation in blood and crooked toes (%). Each coloured line represents the relationship between breeding values for each trait within a year. The broken arrow represents the joint direction of the average breeding value for each trait involved in the trade-off. (From Avendaño *et al.*, 2017)

holds within a year. The data, based on Aviagen breeding programme data, describe the joint trajectory between BWT and OXI, Liveability, LD and CT over 21 years from 1996 to 2017. Each coloured line shows the relationship between the traits estimated breeding values (EBVs) for selection candidates hatched in a specific year. The broken arrow represents the joint direction of the average breeding value for each trait involved in the trade-off. The relationships between traits remain antagonistic within each year but there is a favourable trajectory for each trait because of simultaneous selection; that is, as BWT increases, cardiovascular function, liveability and leg strength increase, while CT decreases. In addition, it can be seen that the worst EBVs for OXI, LIV, LD and CT in 2017 are far better than the maximum of the 1996 birds.

These examples show the principle of multi-trait genetic selection in the presence of trait antagonism and demonstrate that when traits are included in a broad breeding goal and balanced selection is applied, the desired direction in each trait can be and has been achieved.

6.3.3.1 Leg and toe integrity and heart and lung function

Improvements in leg and toe integrity and cardiovascular function trends are confirmed by independent data of Agriculture and Agri-Food Canada (AAFC, 2018), peer-reviewed literature (e.g. leg strengths and toes in Kapell *et al.*, 2012a), the FAWC (2012) report and the EC broiler breeding study (Hiemstra and Ten Napel, 2013). Figure 6.5 shows the Canadian improvements of leg and toe integrity, and heart and lung function, from 1995 to 2018.

Ascites levels as measured by AAFC have fallen correspondingly from about 36 per 10,000 birds slaughtered (1995) to about 8 (2008), and continue to stay low. Although ascites and SDS are now generally considered to be historical problems, companies continue monitoring and selecting against them. Understanding of the fundamental biological bases for the pulmonary hypertension/ascites syndrome in broilers and novel indicator traits may come from immunological and genomics research (Wideman *et al.*, 2013; Hill *et al.*, 2016). In turkeys, similar trends can be seen when the same selection tools are used.

6.3.3.2 Liveability

Reported liveability increases within lines and crossbreeds vary from 0.2–1.0% per year (Hiemstra and Ten Napel, 2013).

6.3.3.3 Contact dermatitis

During recent decades, welfare research on contact dermatitis shifted gradually from hock burn (measured and selected against since the 1970s) to footpad dermatitis (FPD). Kapell *et al.* (2012b) showed how genetic selection can be effective against FPD in contrasting environments – this has been effectively applied since 2008 with decreased levels at the production level as a result.

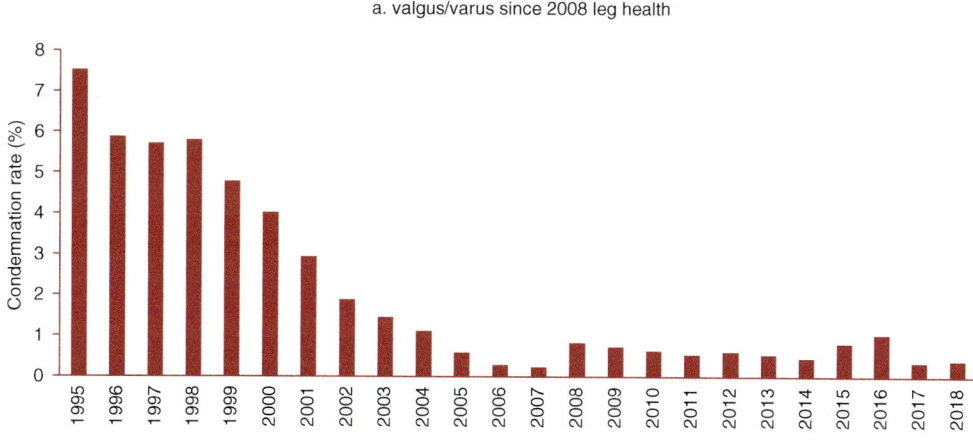

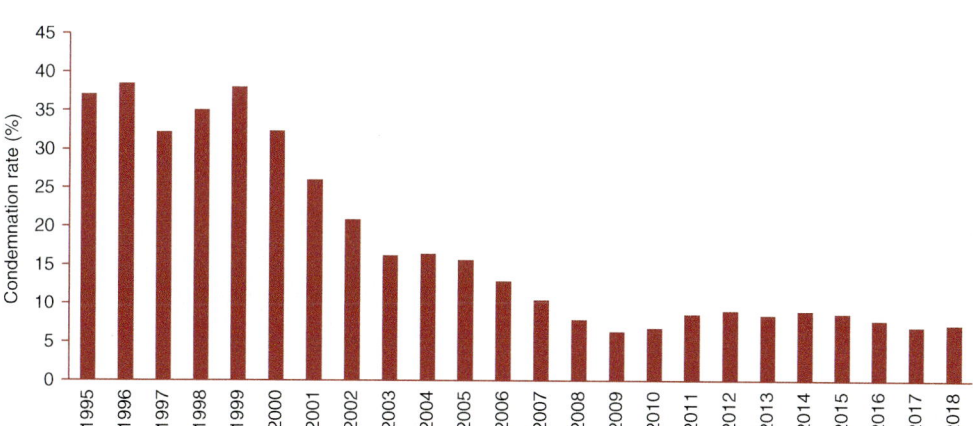

Fig. 6.5. Leg integrity, and heart and lung function trends in Canada (1995–2018). Condemnation rates (birds/10,000 slaughtered) due to: (a) leg integrity – from 1995 to 2007, Agriculture and Agri-Food Canada reports valgus/varus, and from 2008 onwards all leg health condemnations. This explains the slight increase from 2007 to 2008; (b) heart and lung issues expressed as ascites.

6.3.3.4 Body composition

In addition to sustained improvements in biological performance, the modern broiler has shown improvements in body composition. Compared to unselected genotypes, as a function of live weight, the modern broiler has greater leg strength (measured as tibia breaking force) and additional digestive capacity (e.g. greater small intestine surface area, larger pancreas and liver), with no evidence of negative impacts on cardiovascular function, indicating that the modern broiler has a support system suited to handle higher metabolic demands from, and express the genetic potential for, higher biological efficiency and development (Fancher, 2014; Avendaño *et al.*, 2017).

6.3.3.5 Feed intake behaviour

In poultry breeding, recording individual meals during the rearing phase of birds housed in large groups has taken place since 2004, allowing measurements of individual bird intake and feeding behaviour in competition with other birds. These data have shown that there is a significant genetic component in feeding behaviour with heritabilities ranging from 0.24 to 0.57 (Howie *et al.*, 2011). Feeding behaviour in broilers is governed by non-random bouts consistent with periods of hunger and satiety, which are common not only among a wide range of chicken lines but also across selected and unselected species (Tolkamp *et al.*, 2011), meaning that the underlying mechanisms of short-term feeding behaviour are conserved in broiler birds. In addition, short-term feeding behaviour is heritable but independent from selection for FCR, as shown by Howie *et al.* (2011) who found low genetic correlations with feed intake ranging from −0.20 to 0.25. Common feeding behaviour patterns are conserved across chicken lines with widely varying growth rates, and across chickens, turkeys and ducks (Howie *et al.*, 2009, Howie *et al.*, 2010, 2011; Aviagen Group, 2016; Hill *et al.*, 2016).

6.3.3.6 Water intake behaviour

Individual feed intake technology has been expanded to measure individual water intake in both turkeys and broilers. This allows breeding for both feed and water efficiency jointly. Identifying and rejecting birds with very high levels of water consumption for a given level of feed intake will have a positive impact on gut health and the incidence of contact dermatitis through an improvement of the quality of the litter (Aviagen Group, 2016, 2018).

6.3.3.7 Feed intake management of broiler breeders and satiety

An additional item raised is feed intake management of broiler breeders during puberty. The genetic correlation between early and later growth, and similarly between early and late appetite, are very high (Hill *et al.*, 2016). It is possible to manage the chicken growth curve by selection (Ricard, 1975) but at the cost of very large sacrifices in early gain, because of the high genetic correlation with late gain. Dawkins and Layton (2012) argue that by changing selection goals, sampling other populations and incorporating appropriate QTL from non-pedigree populations, the antagonism could be addressed. Crossbred combinations of fast-growing male lines with slower growing female lines have indeed addressed an important part of the issue, with the slower-growing females being fed closer to *ad libitum* levels in both rear and lay, without elevated levels of mortality or multiple ovulation.

Currently, management changes, e.g. feed intake control during puberty, are also a practical and effective approach to manage the antagonism. EFSA (2009) concluded that research in this area is limited and more research was needed to draw firm conclusions about feeding programmes in relation to bird welfare – increased research efforts on alternative feeding systems and their impact on feeding amounts, behaviour and stress indicators (e.g. Dunn *et al.*, 2013; Dixon *et al.*, 2014; van Emous *et al.*, 2014; van Emous, 2015; Chang *et al.*, 2016; Cherrie *et al.*, 2016; Vanderhoydonck, 2016; Lesuisse *et al.*, 2017; Li, 2018; Cherrie *et al.*, 2018) have shown promising results. More research on further fine tuning and better understanding of the various processes is ongoing and is essential to lead to meaningful change.

6.4 Economic Impact of Genetic Improvement

An intrinsic part of the sustainability of the poultry sector is its viability in economic terms and its ability to continue to provide farms and farming for the next generation. Important for this is the resilience of farmers and of the rest of the poultry chain in the short and longer term (Neeteson *et al.*, 2016a). Economic evaluation can give broad direction to breeding strategy but cannot be applied within the selection decisions (Laughlin, 2007).

6.4.1 Welfare improvements and cost savings

Section 6.3 addressed how welfare and production have improved simultaneously over the last two decades, and Fig. 6.5 shows the improvement of valgus/varus and ascites in approximately the same timespan (1995–2018) from AAAF – the AAAF website has similar data on other health traits. Comparisons between selected and unselected or heritage broiler lines provide annual improvement estimates over the last 50 years of around 50 g body weight, feed conversion rate improvements of 15–25 g feed/kg of body weight and around 0.2% higher breast meat yield (Havenstein *et al.*, 1994, Havenstein *et al.*, 2003a, b; Fleming *et al.*, 2007; Mussini, 2012; Zuidhof *et al.*, 2014). Long-term industry data are a reflection of both management practices and genetics – they show improvements of 25–30 g live weight and 16–20 g reduction in feed consumed/kg of live weight, thus indicating sustained improvements in feed efficiency (Laughlin, 2007; Avendaño *et al.*, 2017; Leinonen *et al.*, 2016; NCC, 2018). The economic impact of this is huge as around 70% of total poultry production costs can be attributed to animal feed costs. Low production costs, leanness of the meat and favourable affordability have contributed to making poultry the most used meat by both producers and consumers in the developing world. Developing countries will account for 76% of the output growth, with poultry seeing the most rapid expansion (OECD/FAO, 2018).

6.5 Environmental Impact of Genetic Improvement

As global meat production is expected to increase, the environmental impact of livestock production systems are a key component of sustainable production. The OECD/FAO (2018) forecasts that global meat production is projected to be 15% higher in 2027 relative to the base period 2015–2017.

Animal breeding improves the environmental impact of animal production and sustainable agricultural land use. Poultry is in a favourable position compared to other meats, with values of CO_2 equivalents per unit of edible carcass ranging from around 20–60 for ruminants, 7–20 for pork and 3.7–5 for poultry. Poultry is expected to have a pre-eminent role in satisfying this increasing global demand for meat products with a contribution of 44% of total meat production growth (Williams *et al.*, 2006; Leip *et al.*, 2013; Herrero *et al.*, 2013; Avendaño *et al.*, 2017).

6.5.1   Land savings

Based on 2010, FAO global broiler consumption numbers (86.5 million tonnes, equivalent to 123.6 million tonnes live weight; faostat.fao.org) and annual global commercial field FCR improvements of 0.015 kg/kg (a conservative estimate), cumulative annual savings are $0.015 \times 123.6 = 1.85$ million tonnes of feed. The main chicken meat-producing countries realized a 2010 harvest yield of 466 tonnes of wheat/km^2. The above FCR improvement frees up 1.85 million/466 = about 4000 km^2 of arable land, an area 1.5 times the size of Luxemburg or 3.3 times the size of New York City. This is a cumulative figure realized every year (Neeteson-van Nieuwenhoven *et al.*, 2013).

6.5.2   Life cycle assessment

Avendaño (Avendaño *et al.*, 2017; Avendaño and Corzo, 2018) has calculated the environmental impact of a range of crossbreeds with different average daily gain and FCR using data from crossbreeds in their portfolio. Ross 308 and 708 are globally established commercial broiler crossbreeds; Rowan Range broiler types are slower-growing crossbreeds (eu.aviagen.com/brands/rowan-range). The Ranger genotypes in Table 6.1 fall under the requirements of current EU accreditation schemes like Kip van Morgen and

Table 6.1. Biological performance[1] of seven Aviagen crossbreeds to 2.5 kg body weight. (From Avendaño and Corzo, 2018)

Crossbreeds	ADG	Days	FCRadj	Evis %	Breast %	Liveability %
Ross 308	65	38.5	1.62	73.2	22.7	96.5
Ross 708	63	39.7	1.63	74.1	24.2	97.0
Ranger Classic	49	50.0	1.83	72.2	20.8	97.5
Ranger Premium	50	51.0	1.83	72.7	21.6	97.5
Ranger Gold	47	55.0	1.90	71.4	19.8	97.8
Rowan Ranger	44	62.5	1.99	70.7	18.7	98.0
Rambler Ranger	34	74.6	2.15	69.2	17.0	98.5

ADG = Average Daily Gain; FCRadj = adjusted Feed Conversion Rate to 2.5 kg; Evis% = Eviscerated Yield; Breast% = Breast Yield%.

Beter Leven in The Netherlands, RSPCA Assured in the UK, and Für mehr Tierschutz in Germany.

Biological performance to 2.5 kg BWT difference between the fastest (Ross 308) and slowest (Rambler Ranger) crossbreed is 31.5 g/day for daily gain (ADG), 36.1 days to achieve target weight, 0.53 kg feed/kg BWT, 4% eviscerated yield and 5.7% breast yield. Whereas production differences are large, liveability between the fastest and slowest crossbreed differs more modestly (2%). The reason is balanced breeding goals combining biological performance and liveability and welfare-related traits (Kapell *et al.*, 2012a, Kapell *et al.*, 2012b; Neeteson-van Nieuwenhoven *et al.*, 2013; Avendaño *et al.*, 2017), as is also shown in Section 6.3.

This has huge impact on what is required to produce a kilogram of broiler meat. Annual requirements for feed, agricultural land, water and housing for an integration processing one million birds per week have increased resource need when moving from standard to slower-growing crossbreeds. Compared to Ross 308, the Rambler Ranger requires 32.7% more feed. Water and land increases by 32.7%, which is in direct relationship to the FCR increase (Table 6.1) from 1.62 to 2.15. Housing requirements are three times as high, including the different stocking densities (s.d.) and number of annual cycles which are a function of the number of days to achieve the target weight. s.d.s used were 42 kg/m (Avendaño *et al.*, 2017) for Ross 308 and 708, 38 kg/m² for Rowan Classic and Rowan Premium (i.e. conforming to the Kip van Morgen requirement) and 25 kg/m² for Ranger Gold, Rowan Ranger and Rambler Ranger (i.e. conforming to Beter Leven and RSPCA Assured requirements). Increased resource requirements will impact cost of production; hence, to be profitable, the use of slow-grow options will likely need to be accompanied by capturing premium product prices exceeding the increased resource requirement and cost to remain profitable. In addition, a significant shift towards slow-grow genotypes globally will likely mean an increased pressure on the ingredients market due to poorer FCR, with related higher prices for maize, soya and wheat; hence further driving up the production cost of meat.

With a Life Cycle Assessment (LCA) tool developed by Cranfield and Newcastle University (Poultry LCA Model Version 1.0), Avendaño *et al.* (2017) compared the environmental impact of the broiler crossbreeds in Table 6.1. The 'cradle to farm gate' LCA approach accounted for all inputs and outputs of a production system as described by Leinonen *et al.* (2012). FCR differences between crossbreeds stem from feed amount differences to reach the 2.5 kg target weight at the corresponding age for their target daily gain, not from different feed densities. Input requirements for each system (electricity, heating, water, bedding) were modelled based on Leinonen *et al.* (2012) to 39 days and adjusted to the corresponding age to reach 2.5 kg, meaning input requirements increased linearly with bird age. In this example we have estimated the real environmental impact of breeds and the systems in which they are used. The method can be used for other combinations of breeds and management systems and calculates their environmental impact outcomes.

In their paper, Avendaño *et al.* (2017) show biological efficiency is a major driver for environmental impact differences. As average daily gain and breast percentage increase, global warming potential (GWP) decreases linearly – the opposite trend is seen for FCR. The more biologically efficient crossbreeds have the lowest environmental impact in terms of pollutant emissions and other contributors in LCA. These results are consistent with the findings by Leinonen *et al.* (2012): a free-range and organic production system had a predicted higher GWP of 16% and 28%, respectively, over a standard production system. Acidification potential (AP), eutrophication potential (EP), primary energy use (PEU) and biological performance show similar linear relationships (results not shown). This is in line with Herrero *et al.* (2013), concluding that feed efficiency is a key driver of productivity, resource use and greenhouse gas emissions.

Figure 6.6 visualizes relative GWP, AP, EP and PEU levels with Ross 308 as base reference level (1.0) for the seven crossbreeds as evaluated by Avendaño *et al.* (2017). Ross 308 and 708 showed the lowest predicted environmental impact – predicted increased environmental impact of slow-grow crossbreeds range from around 10% to 30% for GWP and 12–45% for AP, EP and PEU.

As with GWP, these results are in line with Leinonen *et al.* (2012): a standard production system had lower EP, AP and PEU than a free-range and an organic production system. Leinonen *et al.* (2012) showed that an organic production system had environmental impacts with 40% higher EP, 96% higher AP and 59% higher PEU compared to a standard production system. Leinonen *et al.* (2012) also illustrate how biological efficiency, in terms of feed requirements and length of the production cycle, are linked to environmental

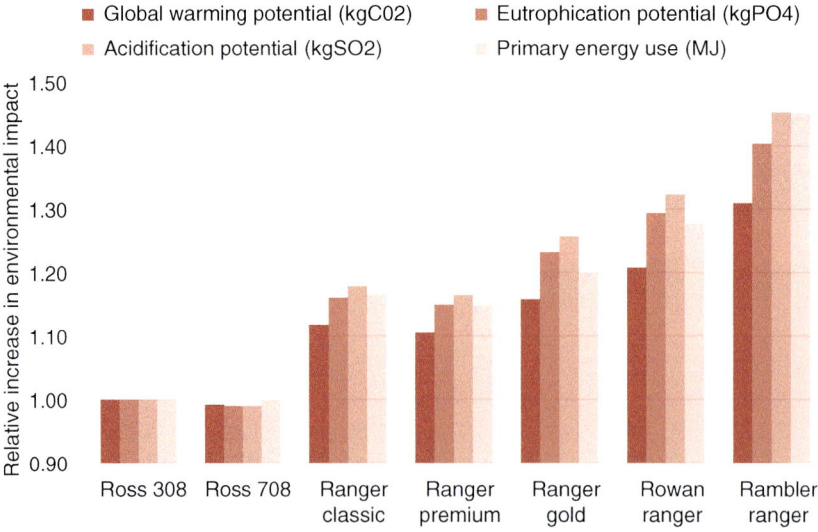

Fig. 6.6. Relative environmental impact of seven Aviagen crossbreeds to 2.5 kg body weight (Ross 308 is the base for comparison at 1.0). (From Avendaño *et al.*, 2017)

impacts of different production system inputs. Feed and water usage had the highest environmental impacts on GWP and PEU (including production, processing and transport) to about 70% of GWP and 65–80% of PEU depending on production system. Farm gas and oil had the second-highest impact in PEU from 12% to 25%, followed by farm electricity (ventilation, feeding and lighting). The use of gas, oil and electricity is generally lower in free-range and organic systems but the lower usage of these inputs does not compensate for the greater FCR and production cycle length.

Summarizing, the impact of crossbreeds with lower biological efficiency on resource utilization and environmental burdens is 30–40% higher compared to conventional genotypes. On the other hand, more broadly, the suitability of a crossbreed to a production system or market segment will depend not only on biological performance but also on consumer preference, product price and other product attributes including the perceived importance between performance and welfare (Avendaño *et al.*, 2017).

6.6 Ethics and Societal Impact of Genetic Improvement

Higher production levels in agriculture have allowed people to take up activities other than food production. Animal-based food has become safer, healthier and more affordable. Food security has improved; and poultry production in particular has contributed to the viability of many poorer regions across the globe (LEAP (Livestock Environmental Assessment and Performance Partnership, 2015).

On the one hand, consumers are more distant from animal food production, as fewer people produce food themselves, or have family members who are farming. At the same time, people are more interested in where their food comes from, and social media can bring the final consumer very close to the food product, the food production system and even the breeding programme, as the customer is more interested and inquisitive as to where meat comes from and under what circumstances it was produced. Farm animal welfare is a major item in the public debate, particularly in affluent regions.

6.6.1 Definitions of animal welfare

A generally used means to describe animal welfare centres around the Five Freedoms, which were developed by FAWC (Farm Animal Welfare Council) (1979) as 'the ideal state to strive for', and the Three Essentials of Stockmanship (knowledge of animal husbandry; skills in animal husbandry; and personal qualities, including affinity and empathy with animals, dedication and patience (FAWC, 2007)). 'This provides a working definition for practical animal breeding centring around the animal's homeostatic balance between (i) its intrinsic potential and requirements; and (ii) its production environment, which includes nutrition, housing, health, the social environment and stockmanship (each to support one of the Five Freedoms of the Farm

Animal Welfare Council). Based on this, breeding goals can be designed to develop appropriate animals for various production systems supported by appropriate management' (Neeteson-van Nieuwenhoven *et al.*, 2013).

6.6.2   Welfare perception by the animal and by humans

People do not all have the same concept of animal welfare, and in addition to physical aspects, people may put more emphasis on mental aspects and on naturalness. However, welfare perception sometimes is related to how a human would perceive a situation or condition. According to Dawkins (2012), it is dangerous to determine animal welfare in terms of anthropomorphism: '[A] lot of people think that good welfare is when animals are allowed to perform natural behaviour, and you can judge welfare by how natural it is … Animals in the wild are regularly chased by predators, and that would be natural.' She indicates that we are ignorant about consciousness – it is better to face this than to pretend we have solved what is consciousness and use that as a basis for animal welfare. Dawkins defines animal welfare as animals that are healthy and have what they want (Rollin, 1993; Fraser *et al.*, 1997; Dawkins, 2012; Neeteson and Avendaño, 2016b).

6.6.3   The food and goodness box

In addition to 'naturalness', other items relevant for public awareness can become part of a general 'food and goodness box', such as dietary health (e.g. consumption of lean animal protein), indoor versus outdoor farming, farm size, use of machinery and technology, feed type (e.g. with or without animal protein, whether or not organic, whether or not genetically modified) and bird health (e.g. use of prophylactic or curative antibiotics; Sonntag *et al.*, 2018). In this general 'good feeling food box', often the separate attributes are not recognized or distinguished: 'animal welfare', 'natural', 'healthy', 'clean', 'whole', 'organic', 'unprocessed', 'freshly made', 'preservative-free', 'foods avoiding artificial ingredients and highly processed foods', and regularly also relate to different types of elimination diets, such as vegan, dairy-free or gluten-free (Mantucci, 2018).

This is in line with the findings of a study by Deloitte (Ringquist *et al.*, 2015) in the USA identifying a shifting of consumer value drivers. While, historically, consumers made decisions based on 'traditional drivers' (price, convenience, taste), a new set of 'evolving drivers', including Health and Wellness, Safety, Social Impact, Experience and Transparency, are now influencing consumer purchasing decisions. This study showed that about 50% of the purchase decisions were influenced by evolving drivers, and the corresponding broader purchase consideration arising from these new drivers was independent from region, age and income level. In addition, the consideration of evolving drivers in purchase decisions was important in both elaborated products (e.g. freshly prepared meals, 66%) and non-elaborated products (e.g. meat, fish and poultry, 49% and dairy, 42%). Clearly, some of the evolving drivers will be more related to retail aspects (e.g. safety), but social impact and health and wellness have a direct link with breeding goals through sustainability, animal welfare and reduced usage of antimicrobials (Avendaño *et al.*, 2017). Spiller *et al.* (2016) highlight that communication based on shared values, i.e. producing food to answer consumer needs, might be a good way forward for animal production as the basic attitude toward farmers in society is positive (Liebert, 2010). The same may be true for the development of products and crossbreeds.

6.6.4 Consumer choice

In addition, more affluent consumers appreciate having an influence on and making a choice from products available to them. These developments are partly the basis for the development of Speciality Production Brands in retail and food service, but also for breeds suited for their various requirements, e.g. colour variants, growth rate, outdoor access appreciative birds. In this way, welfare, as perceived by humans, often as part of a 'food perception box', becomes part of breeding goals for speciality types of production.

6.6.5 Intrinsic conflicts

There may be intrinsic conflicts in the various aspects the public would appreciate. Slower-growing chickens (perceived to have better welfare) have a higher environmental impact. Outside production systems have a higher disease risk, which is in contradiction with food safety and human health. People prefer birds not to be beak-trimmed for welfare reasons, but layers injuring each other by pecking seriously impacts their welfare, even fatally. The transparency role of the breeding sector is, therefore, to provide fact-based information about the various characteristics of crossbreeds under certain production circumstances and to improve crossbreeds for the various markets such that a variety of options are available. A Dutch project, 'Greenwell', is currently developing a model to combine sustainability, welfare and environmental impact.

For each crossbreed, breeders will continue to improve overall welfare, health, productivity and environmental impact, but between the crossbreeds there will be clear differences answering specific demands of concepts and brands. The environmental impact comparison of different growth rate crossbreeds, as highlighted in more detail in Section 6.5, is an example of such information. Other examples are bird colour or eagerness to go outside. Such information allows industry stakeholders (producers, food service, retail) to develop breed/management packages under labels that appeal to consumers in a factual way.

The above-mentioned analysis of consumer drivers means, from a breeding perspective, the following (after Avendaño *et al.*, 2017):

- Continued investment in R&D to elucidate the genetic basis of novel traits and emerging genetic correlations and trade-offs between new and existing traits in the breeding goal. This includes investing in novel recording and selection techniques for the improvement of genetic lines contributing to conventional and slow-grow products; a continual expansion of breeding goals with the ability to change and adapt as a response to current and future changes in global trends. This, in plain words, just means more balanced progress across more traits.
- An expanded genotype portfolio to be able to address the whole spectrum of industry requirements, from covering the sheer scaling up of broiler meat production and global demand in a sustainable way to providing options to emerging niche markets (e.g. free range and organic); thus, the need for large genetic pools suitable to generate broiler products aligned with current and future market needs.
- Continued investments in (i) dialogue and cooperation with the wider stakeholder field, and its developments, so that the future can be designed and implemented; (ii) proof of breeding progress in meaningful terms for stakeholders; and (iii) provision of management support to serve the changing markets.
- In a growing and evolving marketplace, there will be room for both conventional, coloured and slower-growing genotypes. While it is difficult to predict the relative representation of each type of product, globally or regionally, clearly the focus from the primary breeder will be to offer genetic potential suited to all market segments while fulfilling sustainability requirements from economical, biological, welfare and environmental considerations.

References

AAFC (Agriculture and Agri-Food Canada) (2018). Available at: http://www.agr.gc.ca/eng/industry-markets-and-trade/market-information-by-sector/poultry-and-eggs/poultry-and-egg-market-information/condemnations/?id=1384971854399#chicken (accessed 4 October 2018).

Avendaño, S., Watson, K.A. and Kranis, A. (2010) Genomics in poultry breeding: from Utopia to deliverables. *Proceedings of the 9th World Congress on Genetics Applied to Livestock Production*, Leipzig, Germany, Session 07-01.

Avendaño, S., Watson, K.A. and Kranis, A. (2012) Genomics in poultry breeding—into consolidation phases. Proceedings 24th World's Poultry Congress. Salvador, Bahia, Brazil. Available at: http://www.facta.org.br/wpc2012-cd/pdfs/plenary/Santiago_Avendano.pdf (accessed 14 October 2018).

Avendaño, S., Neeteson, A. and Fancher, B. (2017) Broiler Breeding for Sustainability and Welfare – are there Trade Offs? *Proceedings Poultry Beyond 2023 Conference*, Queenstown, New Zealand, p. 17.

Avendaño, S. and Corzo, A. (2018) Selección de Pollos de Engorda - Sostenibilidad y Bienestar Animal. avi News A. *Latina Diciembre*, 2–10.

Aviagen Group (2016) Neeteson, A.M., Swalander, M., Ralph, J. and Koerhuis, A. (eds) *Decades of Welfare and Sustainability Selection at Aviagen. Chickens and Turkeys*, p. 11.

Aviagen Group (2017) Top 5 commitments. Chicken breeding. Available at: http://eu.aviagen. com/assets/Headers/Product-Page-Headers/About-Us/Welfare-Top5Infographic-18-EN.pdf (accessed 20 August 2018).

Aviagen Group (2018) Water utilization in broilers. Aviagen brief. 5pp. Available at: http:// eu.aviagen.com/tech-center/download/1256/AviagenBrief-WaterUtilizationInBroile rs2018-EN.pdf (accessed 31 October 2018).

Bellamy, K. and Bogdan, E. (2016) Dairy and the sustainable development goals. Rabobank Industry Note #574. October 2016, 8pp.

Besbes, B. and Neeteson-van Nieuwenhoven, A.M. (2006) Code of Good Practice for Farm Animal Breeding Organisations. European Poultry Conference. *EPC 2006: 12th European Poultry Conference*, Verona, Italy. World's Poultry Science Association, Beekbergen, The Netherlands, p. 12.

Brundtland report (1987) *Our Common Future. World Commission on Environment and Development*. Oxford University Press, Oxford, UK.

Bryden, W.L. (2016) Water, energy and feed: the trifecta for food security. *In: Proceedings Australian Poultry Science Symposium*, World's Poultry Science Association and Enhancing Poultry Performance. Camperdown, Australia, p. 27.

CBL (Centraal Bureau Levensmiddelenhandel) (2013) Factsheet KIP van Morgen. Available at: http://www.cbl.nl/fileadmin/user_upload/formulieren__factsheets_etc/Factsheet_ Kip_van_morgen.pdf (accessed 14 October 2018).

Chambers, J.R. (1990) Genetics of growth and meat production in chickens. In: Crawford, R.D. (ed.) *Poultry Breeding and Genetics*. Elsevier Science Publishing Co, New York, NY, pp. 599–643.

Chang, A., Halley, J. and Silva, M. (2016) Can feeding the broiler breeder improve chick quality and offspring performance? *Animal Production Science* 56(8), 1254–1262. DOI: 10.1071/ AN15381.

Cherrie, G., Avendaño, S., Dunn, I.C., D'Eath, R.B. and Dixon, L. (2018) *The Effect of Different Feeding Strategies on Female Broiler Breeder Satiety*. International Society for Applied Ethology, Bristol, UK.

Cherrie, G., Avendaño, S., Van Tuijl, O., Dunn, I.C., D'Eath, R.B. *et al.* (2016) The effect of different feeding strategies on broiler breeder satiety. Abstract. UK World's Poultry Science Association, Chester, UK.

Dawkins, M. (2012) What do animals want? A conversation with Marian Stamp Dawkins. The Reality Club. Available at: https://www.edge.org/conversation/marian_stamp_dawkins-what-do-animals-want (accessed 14 October 2018).

Dawkins, M.S. and Layton, R. (2012) Breeding for better welfare: genetic goals for broiler chickens and their parents. *Animal Welfare* 21(2), 147–155. DOI: 10.7120/09627286.21.2.147.

Defra (Department for Environment, Food and Rural Affairs) (2010) Poultry in the United Kingdom. The genetic resources of the national flocks. Available at: https://assets.publishing.service.gov.uk/government/uploads/system/uploads/attachment_data/file/ 69294/pb13451-uk-poultry-faw-101209.pdf (accessed 14 October 2018).

Dixon, L.M., Brocklehurst, S., Sandilands, V., Bateson, M., Tolkamp, B.J. *et al.* (2014) Measuring motivation for appetitive behaviour: food-restricted broiler breeder chickens cross a water barrier to forage in an area of wood shavings without food. *PLoS ONE* 9(7), e102322. DOI: 10.1371/journal.pone.0102322.

Dunn, I.C., Meddle, S.L., Wilson, P.W., Wardle, C.A., Law, A.S. *et al.* (2013) Decreased expression of the satiety signal receptor CCKAR is responsible for increased growth and body weight during the domestication of chickens. *American Journal of Physiology-Endocrinology and Metabolism* 304(9), E909–E921. DOI: 10.1152/ajpendo.00580.2012.

EFFAB (European Forum of Farm Animal Breeders) (2018a) Code of good practice for sustainable farm animal breeding and reproduction. Available at: www.responsiblebreeding.eu (accessed 14 August 2018).

EFFAB (European Forum of Farm Animal Breeders) (2018b) Code-EFABAR the commitment to responsible breeding. Code of good practice for farm animal breeding and reproduction organisations. General brochure. Available at: http://www.responsiblebreeding.eu/uploads/2/3/1/3/23133976/brochure_code_efabar_2014-2016.pdf (accessed 14 October 2018).

EFFAB (European Forum of Farm Animal Breeders) (2018c) Code-EFABAR poultry assured. Available at: http://www.responsiblebreeding.eu/uploads/2/3/1/3/23133976/code_efabar_cepa_folder.pdf (accessed 14 October 2018).

EFSA (European Food Safety Authority) Panel on Animal Health and Welfare (AHAW) (2009) Scientific opinion on welfare aspects of the management and housing of the grand-parent and parent stocks raised and kept for breeding purposes. *EFSA Journal* 8, 1667.

EPB/EFFAB (European Poultry Breeders/European Forum of Farm Animal Breeders) (2009) *The Context of Broiler Breeding*, p. 7.

FABRE TP (Sustainable Farm Animal Breeding and Reproduction Technology Platform) (2008) *Strategic Research Agenda and Annex II: Horizontal Issues*, p. 109.

Fancher, B. (2014) What is the Upper Limit to Commercially Relevant Body Weight in Modern Broilers? *Proceedings of the New Zealand Poultry Beyond 2020 Conference*, Queenstown, New Zealand, p. 20.

FAWC (Farm Animal Welfare Council (1979) Five freedoms. Available at: http://webarchive.nationalarchives.gov.uk/20121007104210/http://www.fawc.org.uk/freedoms.htm (accessed 29 November 2016).

FAWC (Farm Animal Welfare Council) (1979) *FAWC Report on Stockmanship and Farm Animal Welfare*. London, UK. Available at: https://www.gov.uk/government/uploads/system/uploads/attachment_data/file/325176/FAWC_report_on_stockmanship_and_farm_animal_welfare.pdf (accessed 14 October 2018).

FAWC (Farm Animal Welfare Council) (2004) *FAWC report on the welfare implications of animal breeding and breeding technologies in commercial agriculture*. Available at: https://assets.publishing.service.gov.uk/government/uploads/system/uploads/attachment_data/file/325222/FAWC_report_on_the_welfare_implications_of_breeding_and_breeding_technologies.pdf (accessed 14 October 2018).

FAWC (2007) *FAWC Report on Stockmanship and Farm Animal Welfare*. Farm Animal Welfare Council, London.

FAWC (Farm Animal Welfare Committee) (2012) Opinion on the welfare implications of breeding and breeding technologies in commercial livestock agriculture. Department for Environment, Food and Rural Affairs, London, UK. Available at: https://www.gov.uk/government/uploads/system/uploads/attachment_data/file/324658/FAWC_opinion_on_the_welfare_implications_of_breeding_and_breeding_technologies_in_commercial_livestock_agriculture.pdf (accessed 14 October 2018).

Fleming, E.C., Fisher, C. and McAdam, J. (2007) Genetic progress in broiler traits – implications for welfare. *Proceedings of the British Society of Animal Science*, Southport, UK.

Foresight (2011) *The Future of Food and Farming, Final Project Report*. The Government Office for Science, London.

Forni, S., Aguilar, I. and Misztal, I. (2011) Different genomic relationship matrices for single-step analysis using phenotypic, pedigree and genomic information. *Genetics Selection Evolution* 43(1), 1. DOI: 10.1186/1297-9686-43-1.

Fraser, D., Weary, D.M., Pajor, E.A. and Milligan, B.N. (1997) A scientific conception of animal welfare that reflects ethical concerns. *Animal Welfare* 6, 187–205.

Havenstein, G.B., Ferket, P.R., Scheideler, S.E. and Larson, B.T. (1994) Growth, livability, and feed conversion of 1957 vs 1991 broilers when fed "typical" 1957 and 1991 broiler diets. *Poultry Science* 73(12), 1785–1794. DOI: 10.3382/ps.0731785.

Havenstein, G.B., Ferket, P.R. and Qureshi, M.A. (2003a) Growth, livability, and feed conversion of 1957 versus 2001 broilers when fed representative 1957 and 2001 broiler diets. *Poultry Science* 82(10), 1500–1508. DOI: 10.1093/ps/82.10.1500.

Havenstein, G.B., Ferket, P.R. and Qureshi, M.A. (2003b) Carcass composition and yield of 1957 versus 2001 broilers when fed representative 1957 and 2001 broiler diets. *Poultry Science* 82(10), 1509–1518. DOI: 10.1093/ps/82.10.1509.

Herrero, M., Havlík, P., Valin, H., Notenbaert, A., Rufino, M.C. *et al.* (2013) Biomass use, production, feed efficiencies, and greenhouse gas emissions from global livestock systems. *Proceedings of the National Academy of Sciences* 110(52), 20888–20893. DOI: 10.1073/pnas.1308149110.

Hiemstra, S.J. and Ten Napel, J. (2013) *Study of the Impact of Genetic Selection on the Welfare of Chickens Bred and Kept for Meat Production*, DG SANCO 2011/12254. IBF International Consulting, Brussels. Available at: https://ec.europa.eu/food/sites/food/files/animals/docs/aw_practice_farm_broilers_653020_final-report_en.pdf (accessed 14 October 2018).

Herrero, M. and Thornton, P.K. (2013) Livestock and global change: emerging issues for sustainable food systems. *Proceedings of the National Academy of Sciences* 110(52), 20878–20881. DOI: 10.1073/pnas.1321844111.

Hill, W.G. (2016) Is continued genetic improvement of livestock sustainable? *Genetics* 202(3), 877–881. DOI: 10.1534/genetics.115.186650.

Hill, W.G., Wolc, A., O'Sullivan, N.P. and Avendaño, S. (2016) Chapter 11. breeding for sustainability: maintaining and enhancing multi-trait genetic improvement. In: Burton, E., Gatcliffe, J., Masey O'Neill, H. and Scholey, D. (eds) *Poultry Science Symposium Series Vol. 31. UK WPSA symposium September 2014 Chester*. CAB International, Wallingford, UK, pp. 193–213.

Hocking, P.M. (2014) Unexpected consequences of genetic selection in broilers and turkeys: problems and solutions. *British Poultry Science* 55(1), 1–12. DOI: 10.1080/00071668.2014.877692.

Howie, J.A., Tolkamp, B.J., Avendano, S. and Kyriazakis, I. (2009) The structure of feeding behavior in commercial broiler lines selected for different growth rates. *Poultry Science* 88(6), 1143–1150. DOI: 10.3382/ps.2008-00441.

Howie, J.A., Tolkamp, B.J., Bley, T. and Kyriazakis, I. (2010) Short-term feeding behaviour has a similar structure in broilers, turkeys and ducks. *British Poultry Science* 51(6), 714–724. DOI: 10.1080/00071668.2010.528749.

Howie, J.A., Avendano, S., Tolkamp, B.J. and Kyriazakis, I. (2011) Genetic parameters of feeding behavior traits and their relationship with live performance traits in modern broiler lines. *Poultry Science* 90(6), 1197–1205. DOI: 10.3382/ps.2010-01313.

IEC (International Egg Commission) (2018) Egg industry announces commitment to United Nation's Sustainable Development Goals. Available at: https://www.internationalegg.com/egg-industry-announces-commitment-to-united-nations-sustainable-development-goals/ (accessed 26 September 2018).

IPC (International Poultry Council) (2017) IPC adopts sustainability principles. Available at: www.internationalpoultrycouncil.org (accessed 20 August 2018).

IPC (International Poultry Council) (2019) Global poultry leaders sign sustainable development goals commitment with FAO. Available at: www.internationalpoultrycouncil.org (accessed 21 August 2019).

Kapell, D.N.R.G., Hill, W.G., Neeteson, A.-M., McAdam, J., Koerhuis, A.N.M. *et al.* (2012a) Genetic parameters of foot-pad dermatitis and body weight in purebred broiler lines in 2 contrasting environments. *Poultry Science* 91(3), 565–574. DOI: 10.3382/ps.2011-01934.

Kapell, D.N.R.G., Hill, W.G., Neeteson, A.-M., McAdam, J., Koerhuis, A.N.M. *et al.* (2012b) Twenty-five years of selection for improved leg health in purebred broiler lines and underlying genetic parameters. *Poultry Science* 91(12), 3032–3043. DOI: 10.3382/ps.2012-02578.

Laughlin, K.F. (2007) *The Evolution of Genetics, Breeding and Production.* Temperton Fellowship. 15. Harper Adams University, Newport, UK.

LEAP (Livestock Environmental Assessment and Performance Partnership (2015) *Environmental Performance of Poultry Supply Chains.* Version 1. Food and Agriculture Organization, Rome.

Leinonen, I., Williams, A.G. and Kyriazakis, I. (2016) Potential environmental benefits of prospective genetic changes in broiler traits. *Poultry Science* 95(2), 228–236. DOI: 10.3382/ps/pev323.

Leip, A., Weiss, F., Lesschen, J.P. and Westhoek, H. (2013) The nitrogen footprint of food products in the European Union. nitrogen workshop special issue paper. *Journal of Agricultural Science*, 1–14.

Leinonen, I., Williams, A.G., Wiseman, J., Guy, J. and Kyriazakis, I. (2012) Predicting the environmental impacts of chicken systems in the United Kingdom through a life cycle assessment: broiler production systems. *Poultry Science* 91(1), 8–25. DOI: 10.3382/ps.2011-01634.

Lesuisse, J., Li, C., Schallier, S., Leblois, J., Everaert, N. *et al.* (2017) Feeding broiler breeders a reduced balanced protein diet during the rearing and laying period impairs reproductive performance but enhances broiler offspring performance. *Poultry Science* 96(11), 3949–3959. DOI: 10.3382/ps/pex211.

Li, C. (2018) *Transgenerational effects of controlled feed allocation and reduced balanced protein diet on the welfare, cognition and behaviour of broiler breeders.* Doctoraatsproefschrift nr. 1494 aan de faculteit Bio-ingenieurswetenschappen van de KU Leuven, The Netherlands, p. 155.

Liebert, T. (2010) Das image Der Landwirtschaft: ist und Wege zum Soll. Systematische Differenzierungen und kommunikationsstrategische Ableitungen AUS empirischen Befunden. In: Böhm, J., Albersmeier, F. and Spiller, A. (eds) *Die Ernährungswirtschaft im Scheinwerferlicht der Öffentlichkeit.* Lohmar-Köln, pp. 25–46.

Liinamo, A.E. and Neeteson-van Nieuwenhoven, A.M. (2003) Sustainable European Farm Animal Breeding and Reproduction (SEFABAR). *Proceedings. EC ELSA Project 5th Framework Programme for RTD QLG7-CT-2000-01368*, p. 123.

Mantucci, F. (2018) Is 'clean' the new healthy? Mintel blog. Available at: http://www.mintel.com/blog/food-market-news/is-clean-the-new-healthy (accessed 14 October 2018).

McKay, J.C., Barton, N.F., Koerhuis, A.N.M. and McAdam, J. (2000) The challenge of genetic change in the broiler chicken. In: Hill, W.G. (ed.) *The Challenge of Genetic Change in Animal Production.* 27. Occasional Publication British Society of Animal Science, pp. 1–7 DOI: 10.1017/S1463981500040486.

Meuwissen, T.H.E., Hayes, B.J. and Goddard, M.E. (2001) Prediction of total genetic value using genome-wide dense marker maps. *Genetics* 157, 1819–1829.

Mussini, F.J. (2012) Comparative response of different broiler genotypes to dietary nutrient levels. Dissertation. University of Arkansas.

Nature (2018) Animal breeding. Available at: https://www.nature.com/subjects/animal-breeding (accessed 14 August 2018).

Neeteson, A.-M., Appleby, M. and Hogarth, G. (2016a) Making a resilient poultry industry in Europe. Chapter 1. In: Burton, E., Gatcliffe, J., Masey O'Neill, H. and Scholey, D. (eds) *Sustainable Poultry Production in Europe.* CAB International, Wallingford, UK, pp. 3–24.

Neeteson-van Nieuwenhoven, A.-M., Merks, J., Bagnato, A. and Liinamo, A.-E. (2006) Sustainable transparent farm animal breeding and reproduction. *Livestock Science* 103(3), 282–291. DOI: 10.1016/j.livsci.2006.05.016.

Neeteson-van Nieuwenhoven, A.-M., Knap, P. and Avendaño, S. (2013) The role of sustainable commercial pig and poultry breeding for food security. *Animal Frontiers* 3(1), 52–57. DOI: 10.2527/af.2013-0008.

NCC (National Chicken Council) (2018) U.S. broiler performance. Available at: http://www.nationalchickencouncil.org/about-the-industry/statistics/u-s-broiler-performance/ (accessed 14 October 2018).

Neeteson, A.-M., Merks, J., Bagnato, A., Finnocchiaro, R., Sandøe, P. *et al.* (1999) The future developments in farm animal breeding and reproduction and their ethical, legal and consumer implications. *Report EC ELSA Project 4th Framework Programme for RTD BIO4-CT98-0055*, p. 128.

Neeteson, A.-M. and Avendaño, S. (2016b) Sustainability and Productivity. Poultry Meat Production. *Proceedings FACTA Conference 2016*, Expo Dom Pedro, Campinas, SP, Brazil. APINCO Foundation for Poultry Science and Technology, p. 12.

OECD/FAO (Organisation for Economic Co-operation and Development; Food and Agriculture Organisation) (2018) Agricultural outlook 2018-2027. OECD Publishing, Paris. Available at: http://www.fao.org/publications/oecd-fao-agricultural-outlook/2018-2027/en/ (accessed 31 October 2018).

OIE (World Organisation for Animal Health) (2017) *OIE situation report avian influenza.* Available at: http://www.oie.int/fileadmin/Home/eng/Animal_Health_in_the_World/docs/pdf/OIE_AI_situation_report/OIE_SituationReport_AI_6_8May2017.pdf (accessed 26 November 2018).

Rauw, W.M., Kanis, E., Noordhuizen-Stassen, E.N. and Grommers, F.J. (1998) Undesirable side effects of selection for high production efficiency in farm animals: a review. *Livestock Production Science* 56(1), 15–33. DOI: 10.1016/S0301-6226(98)00147-X.

Ricard, F.H. (1975) A trial of selecting chickens on their growth curve pattern experimental design and 1st General results. *Annales de Génétique et de Séléction Animale* 7, 427–444.

Ringquist, J., Phillips, T., Renner, B., Sides, R., Stuart, K. *et al.* (2015) *Capitalizing on the Shifting Consumer Food Value Equation.* Deloitte.

Rodenburg, T.B. and Turner, S.P. (2012) The role of breeding and genetics in the welfare of farm animals. *Animal Frontiers* 2(3), 16–21. DOI: 10.2527/af.2012-0044.

Rollin, B.E. (1993) Animal production and the new social ethic for animals. In: Baumgardt, B. and Gray, H.G. (eds) *Food Animal Well-Being.* Purdue University, West Lafayette, USA, pp. 3–13.

Sonntag, W.I., Spiller, A. and Von Meyer-Höfer, M. (2018) Discussing modern poultry farming systems—insights into citizen's lay theories. *Poultry Science* 0, 1–8.

Spiller, A., Von Meyer-Höfer, M. and Sonntag, W. (2016) *Gibt es eine Zukunft für die moderne konventionelle Tierhaltung in Nordwesteuropa?* No. 1608. Diskussionspapiere, Department für Agrarökonomie und Rurale Entwicklung, Econstor.

Tolkamp, B.J., Allcroft, D.J., Barrio, J.P., Bley, T.A.G., Howie, J.A. *et al.* (2011) The temporal structure of feeding behavior. *American Journal of Physiology-Regulatory, Integrative and Comparative Physiology* 301(2), R378–R393. DOI: 10.1152/ajpregu.00661.2010.

United Nations (2015) Transforming our world: the 2030 agenda for sustainable development. General assembly. 17th session agenda items 15 and 116. A/RES/70/1. Available at: http://www.un.org/ga/search/view_doc.asp?symbol=A/RES/70/1&Lang=E (accessed 14 August 2018).

United Nations General Assembly (2005) *Resolution adopted by the General Assembly 60/1 10. World Summit Outcome 1. Values and Principles.* United Nations, New York.

USPEA (US Poultry and Egg Association) (2018) Ryan Bennett to lead US-RSPE and IPWA. Available at: http://www.uspoultry.org/mediacenter/docs/2018_USRSPE_IPWA_RBennett.pdf (accessed 14 October 2018).

van Emous, R.A., Kwakkel, R., van Krimpen, M. and Hendriks, W. (2014) Effects of growth pattern and dietary protein level during rearing on feed intake, eating time, eating rate, behavior, plasma corticosterone concentration, and feather cover in broiler breeder females during the rearing and laying period. *Applied Animal Behaviour Science* 150, 44–54. DOI: 10.1016/j.applanim.2013.10.005.

van Emous, R.A. (2015) Body composition and reproduction in broiler breeders: impact of feeding strategies. Thesis. Wageningen University – Graduate School of Wageningen Institute of Animal Sciences (WIAS), p. 173.

Vanderhoydonck, C. (2016) Transgenerationele beïnvloeding van de performantie van vleeskuikenmoederdieren en vleeskuikens als gevolg van een aangepaste voederstrategie bij de vleeskuiken(groot)moederdieren (Transgenerational effect on the performance of breeders and broilers as a result of an adapted feeding strategy of their (grand)parents). KU Leuven Bio-ingenieurswetenschappen. Master proef. 134 pp.

Wei, M. and Van der Steen, H.A.M. (1991) Comparison of reciprocal recurrent selection with pure-line selectionsystems in animal breeding (a review). *Animal breeding abstracts* 59, 281–298.

Wideman, R.F., Rhoads, D.D., Erf, G.F. and Anthony, N.B. (2013) Pulmonary arterial hypertension (ascites syndrome) in broilers: a review. *Poultry Science* 92(1), 64–83. DOI: 10.3382/ps.2012-02745.

Williams, A.G., Audsley, E. and Sandars, D.L. (2006) *Determining the Environmental Burdens and Resource Use in the Production of Agricultural and Horticultural Commodities*. Main Report. Defra Research Project IS0205. Cranfield University, Bedford, UK.

Zuidhof, M.J., Schneider, B.L., Carney, V.L., Korver, D.R. and Robinson, F.E. (2014) Growth, efficiency, and yield of commercial broilers from 1957, 1978, and 2005. *Poultry Science* 93(12), 2970–2982. DOI: 10.3382/ps.2014-04291.

7

EU Regulations and the Current Position of Animal Welfare

DONALD M. BROOM*

Department of Veterinary Medicine, University of Cambridge

Summary

In most countries of the world, sustainability issues are viewed by the public as of increasing importance and animal welfare is perceived to be both a public good and a key aspect of these issues. European Union animal welfare policy and legislation on animal welfare has helped animals, has had much positive influence in the world and has improved the public image of the EU. Health is a key part of welfare and the one-health and one-welfare approaches emphasize that these terms mean the same for humans and non-humans. The animals that humans use are described as sentient beings in EU legislation. Scientific information about animal welfare, like that produced by EFSA, is used in the formulation of the wide range of EU animal welfare laws. The European Commission has an animal welfare strategy including the Animal Welfare Platform. However, most kinds of animals kept in the EU are not covered by legislation, and they are subject to some of the worst animal welfare problems, so a general animal welfare law and specific laws on several species are needed. Animal sentience and welfare should be mentioned, using accurate scientific terminology, in many trade-related laws as well as in animal-specific laws.

7.1 Introduction: Sustainability, Welfare as a Public Good and One Welfare

The public in most countries of the world now regard sustainability issues as being of increasing importance. Where resources are exploited, the fact that

*dmb16@cam.ac.uk

© CAB International 2020. *The Economics of Farm Animal Welfare: Theory, Evidence and Policy* (eds B. Vosough Ahmadi *et al.*)

147

something is profitable and there is a demand for the product have become insufficient reasons for the continuation of production. 'A system or procedure is sustainable if it is acceptable now and if its expected future effects are acceptable, in particular in relation to resource availability, consequences of functioning and morality of action' (Broom, 2001a, 2014). The general public, for example those who might buy a product, determine what is or is not acceptable. As explained by Broom (2010):

> [A] system might be unsustainable for several possible reasons. For animal usage systems, including those with an animal production aim, examples of such reasons are: (i) because it involves so much depletion of a resource that this will become unavailable to the system; (ii) because a product of the system accumulates to a degree that prevents the functioning of the system; or (iii) because members of the public find an action that is part of the system, or a consequence of the system, is unacceptable. Where there is depletion of a resource or accumulation of a product, the level at which this is unacceptable, and hence the point at which the system is unsustainable, is usually considerably lower than that at which the production system itself fails.

There are other reasons for unacceptability: harm to the perpetrator, harm to other people, harm to the environment or harm to other animals. Poor welfare of animals has become more important as a reason for system's unacceptability in recent years (Broom, 2017).

For very many people, in Europe and around the world, animal welfare is considered to be a public good (McInerney, 2004) and an important part of sustainability. Conferences involving veterinarians, animal scientists, farmers and retailers are now more likely to include discussion of sustainability issues, including animal welfare, than they were 10 years ago. According to a Eurobarometer survey (2016), awareness of animal welfare continues to increase with 94% of Europeans agreeing on the importance of protecting farm animal welfare. The vast majority of respondents believed that farm animal and companion animal welfare in Europe should be better than it is now.

Health and welfare are terms that apply to humans and all other animals. More and more research shows that human disciplines in these areas are extremely similar to non-human disciplines. The One Health concept is generally defined as a worldwide strategy for expanding interdisciplinary collaborations and communications in all aspects of healthcare for humans, animals and the environment. A resolution promoting the similarity of human and non-human animal health and the need for collaboration between the human medical and veterinary researchers and practitioners was adopted in 2007 by the American Medical Association and the American Veterinary Medical Association. The concept is further explained by Monath *et al.* (2010) and Karesh (2014).

'The welfare of an individual is its state as regards its attempts to cope with its environment' (Broom, 1986). Feelings are a part of coping mechanisms and health refers to coping with pathology, so both feelings and health are important parts of welfare. The term 'welfare' refers to animals, including humans, but not to plants or inanimate objects. The positive or negative state of an individual can be measured scientifically so welfare varies over a range from very good to very poor (Broom and Fraser, 2015).

The similarities, in humans and a range of other species, of studies of stress and welfare referring to physiological, immunological and clinical research were emphasized at a Dahlem conference 'Coping with Challenge: Welfare in Animals including Humans' (Broom, 2001b). It was clear that human psychiatry and medicine could learn from farm animal and other welfare research, and vice versa. More recently, the One Welfare approach has been presented using some parallel arguments to those used in relation to One Health (Colonius and Earley, 2013; García Pinillos *et al.*, 2015, 2016; García Pinillos, 2018; Broom, 2017). People empathize with the suffering of other humans and animals of other species and there are direct links between current welfare and future functioning. When the welfare of individual humans or non-human animals is poor, there is increased susceptibility to disease. As a consequence, improving welfare generally reduces disease. Preventing antimicrobial resistance is good for animal welfare and improved welfare can reduce the need for the use of antimicrobial products. Those with a medical background and those with a veterinary or other biological background benefit from exchanging information, in particular because of the similarities in disease and other causes of poor welfare in humans and other species. Care for people and care for animals used by people is generally better if all are considered as individuals. While this might have seemed too difficult in the past for farmers with thousands of animals, the use of modern precision farming methods currently allows for better responses to individuals in large groups.

7.2   Regulation and Sentience

The EU has some of the highest regulatory standards on animal welfare in the world and the Lisbon Treaty (EU, 2007) recognizes the animals that we use – farm animals, pets, working animals, laboratory animals, animals used for human entertainment – as sentient beings. As a consequence, none of these animals should be considered to be merely property and all legislation should cause them to be treated as sentient individuals. An example of a statement to this effect in EU legislation is in Article 12 of Directive 2010/63 'On the protection of animals used for scientific purposes'. This states:

> Animals have an intrinsic value which must be respected. There are also the ethical concerns of the general public as regards the use of animals in procedures. Therefore, animals should always be treated as sentient creatures and their use in procedures should be restricted to areas which may ultimately benefit human or animal health, or the environment. The use of animals for scientific or educational purposes should therefore only be considered where a non-animal alternative is unavailable. Use of animals for scientific procedures in other areas under the competence of the Union should be prohibited.

The term sentience has generally been used for individuals that have the capacity to have feelings (DeGrazia, 1996; Kirkwood, 2006; Broom, 2006). This is principally a brain capacity so 'sentience means having the awareness and cognitive ability necessary to have feelings' (Broom, 2014). It does not mean actually having the feelings.

> A definition of a sentient being in terms of such abilities is: one that has some ability to evaluate the actions of others in relation to itself and third parties, to remember some of its own actions and their consequences, to assess risks and benefits, to have some degree of awareness and hence to have some feelings. (Broom, 2006, 2014)

There is much evidence for substantial cognitive ability, sophisticated awareness and a range of emotions in the mammals, birds and fish that are used by humans. All vertebrate animals, cephalopod molluscs and decapod crustaceans are considered to be sentient (Broom, 2014). As a consequence, these animals should be protected, or at least their sentience considered, by EU legislation. At present, some farm and laboratory animals are protected but others are not (Broom, 2017). Some EU legislation has not been written or modified to take account of the wording of the Lisbon Treaty. Examples of legislation in which reference to sentient animals is not taken into account include Regulation (EU) 576/2013 on the non-commercial movement of pet animals and Regulation (EU) 1143/2014 on the prevention and management of the introduction and spread of invasive alien species.

7.3   EU Legislation and Other Policies Affecting Animal Welfare

The section of the European Commission that is principally concerned with legislation and other activity in relation to animal welfare is D.G. Santé (Health and Food Safety). However, other sections also take account of animal welfare in the course of their tasks. For example, D.G. Environment has a small team that does this and there are individuals in D.G. Agriculture and Rural Development, D.G. Growth, D.G. Trade and other sections who ensure that animal welfare is considered when they are preparing policy or legislation.

Some legislation and codes of practice are deliberately designed to have an impact on animal welfare but other legislation may have an indirect effect on welfare. Any measure which reduces disease will improve welfare. Some conservation measures lead to fewer attempts to kill animals and hence reduce the poor welfare that often occurs during such attempts. There may be multiple reasons for introducing a legislative measure. Some measures whose aim is to promote sustainable systems may increase biodiversity, reduce pollution and improve animal welfare.

Some of the EU measures with consequences beneficial to animal welfare are listed in Table 7.1. The first EU legislation whose direct aim was to minimize poor welfare in non-human animals was that requiring

Table 7.1. Some EU directives and regulations relevant to animal welfare. (Adapted from Broom, 2017)

Directive 74/577/EEC	Stunning before slaughter
Directive 78/1027/EEC	Veterinary training
Directives 79/409/EEC, 97/49/EC, 2009/147	Conservation of wild birds
Regulations 3626/82, 92/43, 338/97	Wild animals
Directive 83/129/EEC and Regulation 1007/2009	Trade in seal products
Directives 86/113/EEC and 88/166/EEC	Laying down minimum standards for the protection of laying hens kept in battery cages
Directive 99/74/EC	Laying down minimum standards for the protection of laying hens
Directives 86/609/EEC and 2010/63/EU	Protection of animals used for scientific purposes
Directive 76/768/EC and Regulation 1223/2009	Cosmetic products
Directives 88/320/EEC and 99/12/EC	Inspection and verification of good laboratory practice
Regulations 1906/90, 1907/90, 1538/91 and Directive 1999/74/EC	Marketing standards for eggs
Directives 90/425, 91/496, 91/628, 95/29, Regulations 411/98, 1/2005	Protection of animals during transport
Regulation (EEC) 3254/91	Prohibiting the use of leghold traps (imports: humane trapping standards)
Directives 91/629/EEC, 97/2/EC, 2008/119/EC	Laying down minimum standards for the protection of calves
Directives 91/630/EEC, 2001/93/Broom, 2001a/120/EC	Laying down minimum standards for the protection of pigs
Directive 93/119 and Regulation 1099/2009	Protection of animals at the time of slaughter or killing
Regulation 338/97	Protection of species of wild fauna and flora by regulating trade therein
Directive 98/58/EC	Protection of animals kept for farming purposes
Directive 99/22/EC	Keeping of wild animals in zoos
Directive 2007/43/EC	Laying down minimum rules for the welfare of chickens kept for meat production
Regulation (EC) 1523/2007	Banning the placing on the market and the import to, or export from, the Community of cat and dog fur and products containing such fur
Regulation 318/2007	Ban on import of wild-caught birds to be kept in captivity

Continued

Table 7.1. Continued

Directive 2009/156/EC, Commission Regulation (EC) 504/2008, Commission Implementing Regulation (EU) 2015/262, Commission Decision 2000/68/EC, Commission Regulation (EC) 529/2007, Council Regulation (EC) No 21/2004, Commission Decision 2006/968/EC	Marking and tracing animals such as horses, other equids, cattle, sheep and goats (valuable for animal welfare because it reduces disease and opens up possibilities for checking on animals with other welfare problems)
Regulation (EU) 2016/429	Transmissible animal diseases and amending and repealing certain acts in the area of animal health (animal health law) and other legislation aimed at minimizing animal disease and hence improving animal welfare

stunning before slaughter. Soon after this, legislation which specifies the components that must be included in veterinary training was introduced. This prevented the occurrence of poor welfare associated with inadequate training and was beneficial to domestic animals because it required that veterinarians have knowledge about animal welfare. Some poor welfare is also prevented by legislation prohibiting the killing of certain birds and other wild animals. Other measures listed have obvious benefits for animals of the species covered.

Research on animal welfare funded by the EU has led to a large number of publications in the scientific literature. Some EU framework programmes for research and development have funded animal welfare research directly while others have funded projects related to animal welfare, such as animal breeding and animal disease. In Framework Programme 6, the valuable Welfare Quality project was funded and in Framework Programme 7 the innovative Animal Welfare Indicators (AWIN) project was funded (e.g. Blokhuis et al., 2010; Dalla Costa et al., 2014). Under Horizon 2020, there are research topics relevant to animal welfare such as 'breeding livestock for resilience and efficiency' (SFS-15-2016-2017), and 'alternative production systems to address anti-microbial drug usage, animal welfare and the impact on health' (SFS-46–2017). In addition to scientific papers, the AWIN project also produced the Animal Welfare Science Hub (http://animalwelfarehub.com/). This provides information on teaching and research in animal welfare.

7.4   Scientific Information Used in Formulation of Laws

In the EU, a key part of the sequence of events that leads to a directive or regulation about animals has been the production of a scientific report by unbiased scientists. The events leading to an EU directive are exemplified by those concerned with the welfare of calves (Broom, 2009). In the 1970s and 1980s, research results provided evidence for serious welfare problems in closely confined calves. In 1988, the Council of Europe Standing Committee on the Protection of Animals Kept for Farming Purposes (TAP) recommendation concerning the welfare of cattle stated that cattle should be able to make all normal movements for grooming, exercise and other behaviours. This is not possible in a small crate. A European Commission report by a group of scientists in 1990 reviewed the evidence that calf welfare was worse in animals confined in small pens and if there was insufficient iron and roughage in the diet. This led to Directive 91/629/EEC laying down minimum standards for the protection of calves. The use of crates of a minimum size was allowed but a report from the EU Scientific Veterinary Committee was required. There was further research on the effects on calf welfare of diet, confinement, individual rearing and space in groups. In 1994–1995, there was much public pressure for action. The 'Report on the welfare of calves' was produced by the EU Scientific Veterinary Committee (1995), and in 1997, Directive 97/2/EC, phasing out the use of veal crates and inadequate diets, was passed. There are many other examples of legislation based on information from scientific reports in the EU and elsewhere. In recent years, the reports are often referred to as opinions when they are actually scientific reviews. A valuable development in many of the outputs of the European Food Safety Authority (EFSA) concerning animal welfare is the inclusion of systematic risk assessments and benefit assessments (Smulders and Algers, 2009; Broom, 2009; Berthe *et al.*, 2012).

Laws should provide guidance, not just a mechanism to punish (Radford, 2001). EU legislation has some guidance incorporated in it. However, since it is based on scientific reports that are publicly available, the reports and opinions can provide guidance to anyone in the world who may need it. The more recent reports and opinions provide advice on what welfare outcome indicators can be valid and useful for inspectors or farmers attempting to assess welfare in order to comply with laws. Scientific reports are one of the ways in which the EU has helped or had influence on other countries in the world. As explained by Broom (2016, 2017), EU legislation and policies have also had effects internationally and the World Trade Organization (WTO) has accepted that animal welfare can be considered to be a part of public morality and hence to fall within the remit of the WTO. Other world organizations with policies about animal welfare include OIE, FAO, OECD and the World Bank. All of these have been influenced by EU policies and some were initiated because of EU policies. EU legislation on the welfare of calves, poultry, pigs, animals during transport, animals during slaughter, laboratory animals, animals treated with growth-promoting drugs, cloned animals and genetically modified animals has influenced policy and laws in other countries, sometimes being copied by them. Pressure from the general public has had much effect on EU animal welfare laws and policies and is now having the same effects in most countries of the world.

7.5   Current and Future EU Activities in Relation to Animal Welfare

The EU Animal Welfare Strategy, covering the period 2012–2015, was pre-pared in order to help to formulate policy. The impact of this strategy has been evaluated for the European Commission, and in 2017 the EU produced a number of publications that focused on sharing best practice and advancing animal welfare. Those included the 'Preparation of best practices on the protection of animals at the time of killing' and the 'Welfare of farmed fish: common practices during transport and slaughter'. Moreover, during 2016, the Animal Welfare Platform was launched and animal transport guides were proposed. In 2019, the Animal Welfare Platform produced useful guidelines on the welfare of equids. The European Commission released reports in 2018: 'on the impact of animal welfare international activities' and 'on the cross-compliance mechanism of the Common Agricultural Policy (CAP)'. The first EU Reference Centre for animal welfare commenced functioning in 2018. A new strategy is now needed.

Societies in many countries are being changed by the raised level of attention that is given to animal welfare and other concerns about non-human animals. The main focus of EU legislation is on the animals in our food chain but many animals are not yet covered by EU laws. Also, the EU still needs regulation on the welfare of companion animals and wildlife. There is no Union-wide harmonization of many of the codes of practice pertaining to farm animal welfare. Neither is there general legislation on high-quality labelling as currently the labelling is applicable only for eggs. Good treatment of animals can improve long-term productivity and limit the use of antimicrobial drugs and painful practices. Authorities in many countries are being driven by public opinion to support: better dissemination of good practices; voluntary involvement of industries; stricter trade regulations; and enhanced exchange of consumer information.

References

Berthe, F., Vannier, P., Have, P., Serratosa, J., Bastino, E. *et al.* (2012) The role of EFSA in assessing and promoting animal health and welfare. *EFSA Journal* 10(10), s1002, 19–27. DOI: 10.2903/j.efsa.2012.s1002.

Blokhuis, H.J., Veissier, I., Miele, M. and Jones, B. (2010) The welfare Quality® project and beyond: Safeguarding farm animal well-being. *Acta Agriculturae Scandinavica, Section A — Animal Science* 60(3), 129–140. DOI: 10.1080/09064702.2010.523480.

Broom, D.M. (1986) Indicators of poor welfare. *British Veterinary Journal* 142(6), 524–526. DOI: 10.1016/0007-1935(86)90109-0.

Broom, D.M. (2001a) The use of the concept animal welfare in European conventions, regulations and directives. *Food Chain* 2001, 148–151.

Broom, D.M. (ed.) (2001b) *Coping with Challenge: Welfare in Animals Including Humans.* Dahlem University Press, Berlin.

Broom, D.M. (2006) The evolution of morality. *Applied Animal Behaviour Science* 100(1–2), 20–28. DOI: 10.1016/j.applanim.2006.04.008.

Broom, D.M. (2009) Animal welfare and legislation. In: Smulders, F.J.M. and Algers, B. (eds) *Welfare of Production Animals: Assessment and Management of Risks.* Wageningen Academic Publishers, Wageningen, The Netherlands, pp. 339–352.

Broom, D.M. (2010) Animal welfare: an aspect of care, sustainability, and food quality required by the public. *Journal of Veterinary Medical Education* 37(1), 83–88. DOI: 10.3138/jvme.37.1.83.

Broom, D.M. (2014) *Sentience and Animal Welfare.* CAB International, Wallingford, UK.

Broom, D.M. and Fraser, A.F. (2015) *Domestic Animal Behaviour and Welfare*, 5th edn. CAB International, Wallingford, UK.

Broom, D.M. (2016) International animal welfare perspectives, including whaling and inhumane seal killing as a public morality issue. In: Cao, D. and White, S. (eds) *Animal Law and Welfare – International Perspectives.* Springer International Publishing, Switzerland, pp. 45–61.

Broom, D.M. (2017) *Animal Welfare in the European Union.* European Parliament Policy Department, Citizen's Rights and Constitutional Affairs, Brussels, p. 75.

Colonius, T.J. and Earley, R.W. (2013) One welfare: a call to develop a broader framework of thought and action. *Journal of the American Veterinary Medical Association* 242(3), 309–310. DOI: 10.2460/javma.242.3.309.

Dalla Costa, E., Minero, M., Lebelt, D., Stucke, D., Canali, E. *et al.* (2014) Development of the horse grimace scale (HGS) as a pain assessment tool in horses undergoing routine castration. *PLoS ONE* 9(3), e92281. DOI: 10.1371/journal.pone.0092281.

DeGrazia, D. (1996) *Taking Animals Seriously.* Cambridge University Press, Cambridge, UK.

EU (2007) *Treaty of Lisbon amending the Treaty on European Union and the the treaty establishing the European Community* (2007/C 306/01). Available at: http://eur-lex.europa.eu/legal-content/EN/TXT/HTML/?uri=CELEX:12007L/TXT&from=EN

EU Scientific Veterinary Committee (1995) *Report on the Welfare of Calves.* European Commission, Brussels.

García Pinillos, R., Appleby, M.C., Scott-Park, F. and Smith, C.W. (2015) One welfare. *The Veterinary Record* 179, 629–630.

García Pinillos, R., Appleby, M.C., Manteca, X., Scott-Park, F., Smith, C. *et al.* (2016) One welfare – a platform for improving human and animal welfare. *Veterinary Record* 179(16), 412–413. DOI: 10.1136/vr.i5470.

García Pinillos, R. (2018) *One Welfare: A Framework to Improve Animal Welfare and Human Well-being.* CAB International, Wallingford, UK, p. 112.

Karesh, W.B. (ed.) (2014) *One Health. O.I.E. Scientific and Technical Review.* 38. O.I.E, Paris.

Kirkwood, J.K. (2006) The distribution of the capacity for sentience in the animal kingdom. In: Turner, J. and D'Silva, J. (eds) *Animals, Ethics and Trade: The Challenge of Animal Sentience.* Compassion in World Farming Trust, Petersfield, pp. 12–26.

McInerney, J. (2004) *Animal Welfare, Economics and Policy. Report on a Study Undertaken for the Farm & Animal Health Economics Division of DEFRA.* 68. UK Government, DEFRA, London.

Monath, T.P., Kahn, L.H. and Kaplan, B. (2010) One health perspective. *ILAR Journal* 51(3), 193–198. DOI: 10.1093/ilar.51.3.193.

Radford, M. (2001) *Animal Welfare Law in Britain: Regulation and Responsibility.* Oxford University Press, Oxford, UK.

Smulders, F.J.M. and Algers, B. (eds) (2009) *Welfare of Production Animals: Assessment and Management of Risks.* Wageningen Academic Publishers, Wageningen, The Netherlands.

8 Animal Welfare Measures and the WTO Post-EC – Seal Products Case: A Renewed Debate and Research Agenda

Carolina Maciel,[1]* and Bettina Bock[2]

[1]Independent consultant on farm animal welfare policies
[2]Wageningen University & Research and Groningen University

Summary

For nearly 20 years trade officials and scholars debated whether a national measure restricting trade on the basis of animal welfare concerns could be deemed compliant with the rules of the World Trade Organization (WTO). In June 2014, the dispute settlement body of the WTO adopted the decision on the EC – Seal Products case confirming that trade-restrictive measures aimed at safeguarding the welfare of animals can be deemed necessary to protect citizens' moral concerns. While this decision provides long-awaited answers and insights, it does not exhaust the debate on the obstacles for justifying animal welfare trade restrictions. This paper provides an overview of controversies surrounding the topic of animal welfare from a WTO perspective and a brief review of the findings from the EC – Seal Products case. In addition, this chapter calls for further research on potential controversies that may rise in relation to trade measures in contexts beyond seal hunting; like, for instance, regulatory divergences over farm animal welfare measures. In doing so, it recommends that future research pays special attention to the potential controversies associated with the use of animal welfare recommendations elaborated by the World Organisation for Animal Health (OIE) (OIE, 2019).

8.1 Introduction

In many parts of the world, there is an increasing societal urge to safeguard the welfare of animals through the enactment of legal requirements pertaining to

---

*Corresponding author: carolina@tmconjur.com.br

the methods used to capture, raise, transport, kill, slaughter and experiment on them. In addressing this societal demand, many governments find themselves in a difficult position because, ideally, requirements for safeguarding the welfare of animals should be equally binding for domestic and imported products. It is only by ensuring that domestic and foreign products follow equivalent requirements that it will be possible to simultaneously uphold the welfare of animals despite their geographical location and prevent local producers from experiencing economic setbacks in relation to products of lower welfare level that are usually less costly.

However, governmental measures that directly or indirectly impact the flow of imported goods and services are subject to compliance with the rules laid out by the multilateral trade agreements administered by the World Trade Organization (WTO). To a large extent, complying with WTO rules requires government to refrain from adopting measures that will place unnecessary restrictions on the trade flow of goods and services. When there is disagreement as to whether a government has complied with the WTO agreements or not, an international dispute can be initiated by one or more WTO Member States. In such cases, the measure will be examined by an independent group of experts (Panel and Appellate Body), and if inconsistencies are found, the measure needs to be modified or fully withdrawn. Failure to bring a measure under compliance with WTO rules ultimately results in trade retaliation from the other members of the WTO.

The question of whether a trade restriction on import products adopted on the grounds of animal welfare requirements could be deemed necessary in view of a WTO compliance assessment remained open for nearly two decades. It was only in June 2014, in the context of a dispute between the European Union (EU), Canada and Norway, that a WTO authoritative answer was given to that question. The answer emerged from the analysis of whether the European ban on seal products could be considered a lawful trade restriction on the grounds of a necessity to protect moral concerns. The dispute, known as EC – Seal Products (WTO, 2015),[1] is the first case to specifically address animal welfare measures under the WTO-covered agreements, and the third case in the history of multilateral trading system to analyse the applicability of the public moral exception. The ruling in this pioneer case was favourable to the interpretation that citizens' concerns over the level of welfare experienced by animals falls within the meaning of the public moral exception. This ruling is a remarkable vindication of the right of governments to justify animal welfare trade restrictions that would otherwise be regarded as a breach of the WTO core obligations.

This chapter reviews the ruling of the EC – Seal Products case with the primary aim of describing how the Panel and Appellate Body (AB) assessed the question of necessity. Understanding the context and the reasoning adopted in their decision is of great relevance given that this case sets an important legal precedent for future assessment of animal welfare measures under the WTO framework. In addition, the chapter explores a new set of further questions that will advance the debate of animal welfare measures and the WTO system after the EC – Seal Products case. This secondary aim is relevant considering that a full import ban, like the one enacted in the case of seal products, might not be deemed necessary by WTO adjudicators in other contexts of animal protection. This might occur if other less trade-restrictive alternatives such as standards and certification schemes are found to be reasonably available. Accordingly, this chapter argues that further debate and research are needed to assess the role and status of the animal welfare recommendations set by the World Organisation for Animal Health (OIE) (OIE, 2019).[2]

8.2   Setting the Scene: The WTO System and the Animal Welfare Measures

The World Trade Organization (WTO) is an international organization established in 1995 after the conclusion of the Uruguay Round of Multilateral Trade Negotiations (1986–1994). The WTO provides the legal and institutional framework for the implementation and monitoring of multilateral and plurilateral agreements covering the regulation of trade in goods, services and intellectual property. Among these agreements are the General Agreement on Tariffs and Trade[3] (GATT), the Agreement on Sanitary and Phytosanitary Measures (SPS), the Agreement on Technical Barriers to Trade (TBT) and the Agreement on Agriculture (AoA). In addition, the WTO provides a framework for member countries to negotiate new trade agreements and to settle disputes arising from the interpretation and application of current agreements. The procedures for adjudicating legal disputes concerning compliance with WTO agreements are defined in the Understanding on Rules and Procedures Governing the Settlement of Disputes (DSU).

Permeating all WTO trade agreements is the principle of non-discrimination that requires Member States to formulate their policies in a manner that is not discriminatory towards foreign goods and services. The non-discrimination principle is grounded on two rules: the most favoured nation (MFN) and the rule of national treatment (NT). In view of the first rule, a Member State is obliged to accord the most favourable tariff and regulatory treatment given to a product of any other member at the time of import or export to all other members importing 'like products'; in view of the second rule, a Member State is obliged to accord imported products treatment no less favourable than that conferred to 'like products' of its national origin. Derogation from non-discrimination rules is permitted under strict conditions as defined by specific provisions known as exceptions. The most important of these exceptions are listed in Article XX of the GATT Agreement.

At the time of the negotiations of the Uruguay Round (1986–1994), animal welfare was not such a prominent and widespread government responsibility as it is nowadays. Even though political debates on whether or not animals are entitled to protection date back to the ancient Greek philosophers, and the first law prohibiting cruelty to animals was enacted in 1822 in England, the foundations of current animal welfare regulation were developed in the late 1960s on the basis of the *Five Freedoms*[4] (Veissier *et al.*, 2008; Maciel, 2015). The concept of the five freedoms encouraged scientific and political debates about the need to complement the existing anti-cruelty regulations with welfare provision that identifies minimum requirements for housing, feeding, transporting and slaughtering of farm animals as well as minimum standards for the capture of wild animals, and defines strict limits for the use of animals in laboratory tests. This understanding of the need to expand the animal protection law emerged first among European countries, and only in more recent years it has been embraced by other countries. It is thus most probably because animal welfare was not widely recognized as a legitimate subject at the time of the negotiations round in the 1990s that earlier attempts failed to include a reference to animal welfare in one of the agreements leading up to the creation of the WTO system.[5]

The absence of an explicit reference to animal welfare in the WTO agreements resulted in a growing disagreement among policy makers, legal scholars and other interested parties on whether or not a trade measure pursuing animal welfare concerns could be considered a breach of WTO obligations. In June 2000, the then European Community (EC) reported[6] that there was a growing concern among consumers, producers and animal protection organizations that while the WTO is working to enhance the framework for the liberalization of international trade, it does not provide legal clarity on how to address animal welfare issues in view of differing levels of animal protection granted by WTO members. The EC highlighted the importance to secure the right of those WTO members that apply high animal welfare standards to maintain their protection. For the EC, the elaboration of a new multilateral agreement specific to animal protection was one possible way to solve the current uncertainties revolving around the lawfulness of restricting trade on the basis of animal welfare concerns. The EC, however, argued that there were also opportunities to address this issue through existing WTO agreements, namely the SPS Agreement, the TBT Agreement, the AoA Article 20 or one of the exceptions listed in Article XX of GATT.

Despite the efforts of the European Commission to drawn attention to the need to address the question of animal welfare within the WTO, the issue did not earn significant space on the negotiation rounds of the WTO. Thus the only option left for seeking clarity on how animal welfare can be accommodated within the existing WTO system was through the text interpretations provided by rulings delivered by the Panel and AB in a context of trade disputes. The rulings adopted in the 1990s[7] also did not offer many grounds to positively conclude that trade restrictions pursuing animal welfare concerns could be deemed a necessary measure under a WTO compliance scrutiny. As a result of the lack of animal welfare language in the WTO agreements and the outcome of early WTO trade disputes, governments became hesitant to adopt animal welfare measures. The *regulatory chilling effect*[8] of the multilateral trading system on the progress of animal welfare regulation is often illustrated with the reluctance demonstrated by the EU regarding the implementation of a ban on the use of leg-hold traps for animals captured for fur, the use of animal testing for cosmetics and the use of battery cages for laying hens. (For further information on these examples see CIWF, 2002; Stevenson, 2002,Stevenson, 2009; and Maciel, 2015.)

Over the years, other trade disputes[9] allowed for some optimism concerning an eventual interpretation by a Panel/AB that a product produced in good animal welfare conditions would not be considered as 'like' products produced in poor animal welfare conditions when examining an alleged breach of the MFN and NT obligations. Nevertheless, it was still very important to have an interpretation from the Panel/AB that a trade restriction based on animal welfare concerns could be found necessary within the scope of GATT's exception of Article XX item (a). Conjectures about it have been around since the early days of the WTO[10] (e.g. Stevenson, 2000; Jenkins and Stumberg, 2001). However, authors assessing the potential success of animal welfare to pass the multilayer scrutiny of a GATT Article XX(a) had different opinions on the outcome (see, for instance, Thomas, 2007 for a positive evaluation, and Fitzgerald, 2011 for a negative one).

8.3   The Analytical Steps of GATT Article XX(a)

The Article XX offers the possibility of derogation from general WTO trade obligations such as the MFN and NT for unilateral trade measures in pursuit of specified purposes. The article is composed by an opening statement known as 'chapeau', which sets some requirements for the manner in which the measure should be applied, plus an exhaustive list of reasons for justifying the exemption such as the need to protect public morals. Below, the text of Article XX(a) is presented *in verbis*:

> Article XX: General Exceptions
>
> Subject to the requirement that such measures are not applied in a manner which would constitute a means of arbitrary or unjustifiable discrimination between countries where the same conditions prevail, or a disguised restriction on international trade, nothing in this Agreement shall be construed to prevent the adoption or enforcement by any contracting party of measures:
>
> (a) necessary to protect public morals;
>
> [...]

The public moral exception has been in place since the original 1947 GATT Agreement, but for almost 30 years it has been dormant. The first dispute concerning the application of this article was in 2009 in a case known as China – Audio visual.[11] Prior to this, the public morality exception had only been invoked in 2004 in the context of the General Agreement on Trade in Services (GATS), in a case known as US-Gambling[12]. In both cases, the justification of non-compliance with WTO rules on the basis of public morality did not fully succeed, as it failed to pass the various analytical steps. The justification of a measure under Article XX(a) requires a sequential process of 'weighting and balancing' a number of distinct factors relating both to the measure sought to be justified as 'necessary' and to possible alternative measures that may be reasonably available to the responding member to achieve its desired objective.

The dynamic of interpretation, ultimately, seeks to determine (i) whether the measure indeed protects a moral value; (ii) whether the measure is deemed necessary to achieve the desired protection; and (iii) whether the measure is applied in a manner that constitutes an arbitrary or unjustified discrimination between countries where the same conditions prevail. Thus, for justifying a WTO-inconsistent measure under Article XX(a), a respondent first needs to establish that its national measure falls within the content and scope of the concept of 'public morals', which has been interpreted as denoting a standard of right and wrong conduct maintained by or on behalf of a community or nation.[13] Next, the defendant must provide proof that the measure is necessary to the achievement of the objective by demonstrating the contribution made by the measure to secure compliance with the relevant law or regulation and that alternative measures that are less trade-restrictive were not reasonably available. The final step in the defence of Article XX involves establishing that the challenged measure complies with the 'chapeau' to prevent the abuse of the exceptions.

As one can see, defending a measure under Article XX(a) is not an easy task. Because of this difficulty, the European attempt to defend its seal regime on the basis of the public morality exception caused a lot of anxiety and unrest. Following Cook and Bowles (2010), the outcome of the first WTO dispute that specifically addressed an animal welfare measure would result in either a vindication or a rebuttal of all the years of great caution on the part of governments to take further steps in restricting the trade of goods produced under poor welfare conditions.

8.4   EC – Seal Products Case: Confirming the *Necessity* of an Import Ban

This dispute around seal products arose after a complaint brought by Canada and Norway against the regulatory scheme established by the EU concerning seal products (Regulation 1007/2009 and 737/2010). In brief, these regulations banned the sale and marketing of seal products[14] in the EU territory on the grounds of a moral outrage concerning the inhumane killing of seals. This general prohibition provided for three exceptions: (i) products derived from non-commercial hunts conducted by Inuit or other indigenous communities (IC exception); (ii) products obtained from hunts conducted for the sole purpose of the sustainable management of marine resources (MRM exception); and (iii) products entering the EU market as items of personal use by travellers for non-commercial reasons (traveller's exception).

Canada and Norway, which are among the biggest producers of seal products, complained that the above measures are inconsistent with the obligations of the European communities under the multilateral trading system. They claimed that the EU Seal Regime constituted a violation of several WTO provisions of the GATT and TBT Agreements; notably the provisions of non-discrimination embodied in the rules on most favoured nation (MFN), national treatment (NT) and the prohibition of quantitative restriction of GATT.

In order to justify its measure, the EC invoked the public morals exception in GATT Article XX(a). The EC argued that the Seal Regime was necessary to protect deep and long-standing moral concerns of the EU public with regards to the presence of seal products on the European market that may have been obtained from animals killed in a way that causes excessive pain, distress, fear or other forms of suffering for the animals.

The assessment of the WTO adjudicatory bodies on the possibility of justifying the seal product ban under Article XX(a) comprised a thorough analysis of (i) the nature of the measure to assess its purview under the notion of public morality; (ii) the necessity of the measure for achieving the objective of protecting EU citizens' moral concerns about the welfare of seals; and (iii) the consistency of the measure with the 'chapeau' of Article XX, which requires a measure not to be applied in a manner that constitutes a means of arbitrary or unjustifiable discrimination between countries where the same conditions prevail.

The Panel report (circulated on 25 November 2013) concluded on the basis of its examination of the text and the legislative history of the EU Seal Regime, as well as other evidences pertaining to its design, structure and operation, that the objective of the EU Seal Regime was indeed to address the moral concerns of the EU public with regards to the welfare of seals. The Appellate Body report (circulated on 22 May 2014) upheld the Panel's finding that the EU Seal Regime is 'necessary to protect public morals' within the meaning of Article XX(a) of the GATT Agreement (1994).

In assessing the necessity of the sales and marketing ban for reaching the desired level of protection of EU citizens' morality, the Panel analysed the contribution of the measure to the fulfilment of the objective in parallel with an analysis of the availability of less restrictive alternative measures. Looking at the facts and arguments presented by both parties, the Panel concluded[15] that the EU Seal Regime is capable of making, and does actually make, an important contribution to the achievement of its stated objective. Through the prohibition of the import and sale of seal products, the EC prevents its citizens, individually and collectively, from being exposed to and participating as consumers in commercial activities that involve the inhumane killing of seals.

The Panel also examined alternative measures identified by the complainants consisting of a market access for seal products conditioned on compliance with animal welfare standards combined with certification and labelling requirements and the defendant's rebuttal that such measures were not an appropriate alternative. The Panel's reasoning considered the potential contribution of the alternative measures for the fulfilment of the EU's aim from the perspective of its feasibility of enforcement and potential risk of products from inhumane killing of seals being exposed to European citizens. In arguing for the feasibility of the alternative measures, Canada and Norway claimed that poor animal welfare risks are 'commonplace'[16] in situations involving the killing of animals, either in other terrestrial wildlife hunts or in slaughterhouses, and so it would not be reasonable to heavily constrain seal hunts. In its rebuttal, the EU explained that, contrary to the complaining party's arguments, the environmental conditions under which seal hunts take place render it impossible to apply and enforce requirements of humane killing methods in an effective and consistent manner and to guarantee the avoidance of pain, distress, fear and other forms of suffering during the killing and skinning of seals.

Upon a careful weighing of circumstances, risks and challenges associated with the hunting of seals, the Panel concluded[17] that although the proposed alternative measures (certification and labelling) are less trade-restrictive, the adoption of an import condition (instead of a full import ban) could not be considered reasonably available taking into account the risks associated with the non-fulfilment of the objective, that is the protection of EU markets from an activity that violates its citizens' morals. Hence, the Panel found that the ban on the importing and marketing of seal products was a necessary measure to ensure the fulfilment of the EU's objective. This conclusion confirms that governmental measures intended to protect and promote the welfare of animals can legitimately result in trade restrictions provided they are not applied in a discriminatory or arbitrary manner.

Although upholding the Panel's understanding that the EU measure was not more trade-restrictive than necessary due to its legitimate objective (public moral concerns on seal welfare), the AB found[18] that the EU had failed in designing legislation that prevented arbitrary discrimination resulting from the exception for hunts conducted in the framework of maritime resource management and the exception related to indigenous communities.[19] Consequently, the EU was asked to modify its regulatory framework regarding those two specific areas to ensure that the manner in which its seal regime is implemented meets the requirements of the chapeau of Article XX. The adjustments carried out by the EC resulted in the elimination of the MRM exception and the modification of the IC exception. The revised EU Seal Regime enacted by the EC in response to the findings of arbitrary discrimination resulted in a much stricter regulatory framework for the protection of seals.

The ruling in this pioneer case of animal welfare sets an important legal precedent upholding the understanding that moral concerns regarding animal welfare provide legitimate reasons for imposing trade restrictions. The fact that the EU Seal Regime partially survived the analytical step of the Article XX(a), and that the revised measure resulted in a stricter market access, is a remarkable achievement that is expected to increase government's confidence in taking further measures to address societal demands for greater protection of animals.

8.5   Beyond Seal Products: Renewing the Debate and Research Agenda

Every ruling of the WTO generates a number of publications, and the EC – Seal Products ruling is no different. Either from an animal welfare perspective or from a broader context, the EC – Seal Products decision has been at the focus of many recent publications. Some authors criticized the decision, arguing that it lacked 'graspable and objective' criteria with regards to the scope of public morality, which could eventually undermine the WTO system by allowing all sorts of trade restrictions to be justified (e.g. Möllenhoff, 2015; Mavroidis, 2015). Others highlighted the rightfulness of the decision to confirm the basis of GATT Article XX(a) to justify animal welfare trade restrictions on the basis of moral concerns (e.g. Howse *et al.*, 2014; Offor and Walter, 2017; Sttuhent, 2017). In doing so, these commentators indicate that WTO is striking a good balance between the regulatory autonomy of WTO members and the multilateral trading system. There have also been some publications that sought to extrapolate the findings and give new insights to other long-lasting debates, for instance the potential of other moral issues such as labour standards to be justified (e.g. Cottier, 2018, or by re-examining related key WTO concepts such as 'treatment no less favourable' (e.g. Du, 2016).

What seems to be missing from the academic debate post-EC – Seal Products is how other animal welfare-related trade restrictions could be assessed under the WTO, and what challenges the responding members might face. This is relevant, as the positive outcome of the Seal case does not provide for all other animal welfare restrictions to be automatically justified, and there are still many questions to answer, such as the likeness between products with higher and lower standards. Thus, continuing the research is important as the regulatory landscape of animal welfare is expanding and eventual new disputes might bring additional challenges to solve. For instance, future disputes might not be between a country with an animal welfare trade measure and a country without it; but it might be a dispute between two countries with different welfare requirements, which brings the challenge of defining what is the acceptable level of stringency on the protection of an animal from a trade law perspective. Taking pigs as an example, one finds that many countries have adopted (or are about to adopt) minimum requirements on housing, feeding, surgical procedures and other handling manners with the objective of protecting the welfare of pigs (Driver, 2017).[20] However, this legislation varies in scope and specificity, which suggests that trading nations have different views on the necessary level of protection required for pigs at the farming stage. Consequently, further research on the use of animal welfare standards from a legal trade perspective is important.

When addressing the question of necessity of the EU Seal Regime containing a full ban on the sale of all products derived or obtained from seals, the Panel and AB initiated a discussion on the possibility of adopting animal welfare standards for the humane killing of seals. The discussion aimed at assessing whether a market access for seal products conditioned on compliance with requirements for a humane killing of seals could be considered a reasonable available alternative measure to address the EU's moral concern over the welfare of seals. The Panel observed[21] a discrepancy between experts and other sources as to what would constitute adequate welfare standards and criteria,[22] which resulted in difficulties in asserting what a humane killing method adapted to the conditions of seal hunt would be. More generally, the Panel detected that the discrepancy of opinion in relation to animal welfare requirements and standards stemmed in part from 'differing assessment and tolerance of the risks involved in the application and monitoring of killing methods',[23] and thus 'requirements under the alternative measure could possibly span a range of different levels of stringency or leniency'.[24]

Accordingly, the Panel acknowledged that 'certification schemes of greater specificity and rigour may be considered less reasonably available to the extent that they would require greater expenditure and practical challenges of implementation. At the same time, schemes that are not designed to account for the actual welfare outcomes of the seals from which products are derived may be considered comparatively more reasonably available.'[25] In view of these hypothetical versions of alternative measures and absence of 'clearly articulated standards from the complainants',[26] combined with the evidence of the risks and challenges inherent in the activity of seal hunting, the Panel concluded that the proposed alternative measure (conditioned market access) was not reasonably available. Hence the EU's full sales ban on all seal products was not more trade-restrictive than necessary to fulfil its objective of preventing participation of the EU public as consumers of products deriving from the inhumane killing of seals. This is an important finding in the light of the increasing variety in requirements defined in animal welfare regulations across the globe, which probably also results from the same lack of unanimity that the Panel referred to in the analysis of the EU – Seal case.

This chapter claims that it is high time to engage in a further in-depth analysis of the role and status of the various animal welfare standards, but in particular the ones elaborated by the World Organisation for Animal Health (OIE) (OIE, 2019). This is an intergovernmental organization created in 1924, and which since 2004 has elaborated 14 sets of welfare recommendations for terrestrial animals and four for aquatic animals. These recommendations cover a variety of species and situations including the production, transportation, slaughter and killing of farm animals. The elaboration of animal welfare recommendations by the OIE is a relatively new development in the global governance landscape of animal protection. Nonetheless, a survey conducted in Brazil has shown that the OIE activities in this policy field have already positively impacted the level of awareness and degree of legitimacy of animal welfare measures among government officials (Maciel, 2015). The growing relevance of OIE engagement on the policy field of animal welfare is catching the attention of all sorts of stakeholders, including non-governmental organizations of animal protection, food companies, production associations and scholars in general. Despite this growing attention, very little can be found in the literature on animal welfare and international trade law about it beyond plain references to the existence of international recommendations set by an intergovernmental body named OIE. This paper argues that a thoughtful consideration of OIE animal welfare recommendations and their possible role in the eventuality of a new trade dispute over animal welfare measures is an essential next step in the research agenda of international law scholarship.

How the OIE has evolved over the years to become the leading reference in animal welfare standards is, in itself, an interesting subject for study. The OIE was created by 28 states to provide coordination among veterinary officials with regard to the prevention and control of contagious livestock diseases such as rinderpest and foot-and-mouth. Over the years, the organization grew in membership,[27] mandate, policy tools and importance within the international landscape. Among the key moments for the organization was the decision in 1968 by the International Committee of the OIE (nowadays known as the World Assembly of National Delegates) to formally adopt a number of recommendations to be used by veterinary authorities of the member countries to set up national policies for the early detection, reporting and control of pathogenic agents, including zoonotic ones, in terrestrial animals (mammals, birds, reptiles and bees) as a means to prevent their spread via the international trade in animals and animal products. These recommendations have been compiled into a document, which later became known as the OIE Animal Health Code[28] (AHC), published annually with revisions and/or additions.[29] Another turning point in OIE history is linked with the establishment of the WTO in 1995, when the AHC gained a whole new status under international law as a result of the SPS Agreement that defined the OIE as the relevant international organization for the development of standards, guidelines and recommendations for animal health and zoonoses.

The WTO SPS Agreement is the multilateral framework of rules and disciplines that guides the development, adoption and enforcement of sanitary and phytosanitary measures, which may, directly or indirectly, affect international trade. In the context of this Agreement, SPS measures can only be applied 'to the extent necessary to protect human, animal or plant life or health' and they must be 'based on scientific principles' and 'not maintained without sufficient scientific evidence'.[30] This agreement encourages the harmonization of SPS measures with relevant international standards, guidelines or recommendations. As such, Article 3.2 of the SPS Agreement indicates that measures conforming to the relevant international standards, guidelines or recommendations 'shall be deemed to be necessary to protect human, animal or plant life or health, and presumed to be consistent with the relevant provisions of this Agreement and of GATT 1994'. Accordingly, if a country designs its national animal health regulations in conformity with the recommendations of the OIE Animal Health Code, its animal health measure is deemed to be WTO-consistent. But if the national measure deviates from the OIE AHC, then the country needs to provide a scientific assessment of risks and a clear determination of the appropriate level of sanitary protection it seeks to achieve, taking into account technical and economic feasibility. Failure to provide a scientific justification for deviating from the OIE recommendation may result in a finding of a breach of WTO obligation in a given trade dispute.[31]

Whereas the roles and status of OIE animal health recommendations are known, doubts remain in relation to the OIE animal welfare recommendations, despite both being laid out in the AHC. These uncertainties result for a number of reasons, including (i) the fact that OIE's work on animal welfare began only after the enactment of the SPS Agreement, which is the document that grants special status to the OIE animal health recommendation within the WTO framework; and (ii) the existence of controversies about whether or not animal welfare is covered by the concept of 'SPS measures'. There are recurring references in literature and other documents that animal welfare is not covered by the SPS Agreement. Despite this, Maciel (2015) argues that it is not impossible for a Panel/AB to decide otherwise given the drafting history of this Agreement and the interpretative approach applied in the EC – Biotech[32] when deciding the applicability of the SPS Agreement on the analysis of regulatory measures concerning the control of genetically modified organisms (GMO).

Beyond clarifying the status of the OIE animal welfare recommendation within WTO SPS Agreement, it is also recommended that further research assesses how OIE animal welfare recommendations can facilitate or hamper a WTO member's justification of its national measure in an eventual trade dispute over the level of stringency. It is the opinion of some stakeholders that OIE animal welfare recommendations are suboptimal choices as they are based on the lowest common denominator (Sharpless, 2008; White, 2013). Thus a country seeking to adopt a measure above the OIE recommendation for animal welfare would need to provide a scientific justification based on risk assessment if the measure were to be examined from an SPS Agreement perspective. In addition, future research shall look at the implications of eventually having farm animal welfare measures examined under the framework of this agreement, given the inherently moral component of animal protection initiatives and the exclusive scientific scope of SPS measures.

Another perspective on studying the role and status of OIE animal welfare recommendations is from the TBT Agreement. The rules and procedures regarding the development, adoption and application of voluntary product standards, mandatory technical regulations, and the procedures for assessing conformity are provided in this Agreement. Similar to the SPS Agreement, TBT also encourages members to base their measures on international standards as a means to facilitate trade. However, the TBT only refers to the concept of 'relevant international standards' without naming any specific organization. It is reasonable to assume that there will be no disagreement in recognizing the OIE recommendation as the relevant international standard on the topic. In contrast, controversies can be expected over the assessment of whether a national measure used the OIE recommendation 'as a basis' for developing its regulation. This may occur given the language applied in the OIE recommendations that are broadly expressed by the use of 'should' instead of 'must' (White, 2013). Whereas the first term connotes an advice, the latter connotes a necessity; and it is precisely the deliberation over the necessity that poses obstacles in the justification of a measure.

It is also worth giving attention to the collaboration between OIE and the International Organization for Standardization (ISO) initiated in 2011, and which resulted in the elaboration of ISO/TS 34700:2016.[33] This is a technical specification on 'animal welfare management -- general requirements and guidance for organizations in the food supply chain'. As the ISO describes, this document provides requirements and guidance for the implementation of the animal welfare principles as described in the introduction to the recommendations for animal welfare of the OIE TAHC (Chapter 7.1). It applies to terrestrial animals bred or kept for the production of food or feed and is limited to aspects for which process or species-specific chapters were available in the OIE TAHC at the time of its publication in December 2016.[34] Furthermore, studies to look at how ISO/TS 34700:2016 has been incorporated into public and private standards and how these developments intersect with the rights and obligations of WTO members are suggestions for future research.

8.6   Conclusion

After almost two decades of uncertainty and controversies over the lawfulness of restricting trade on the basis of animal welfare concerns, the ruling in the EC – Seal Products case confirmed that citizens' concerns over the welfare of animals fall within the content and scope of the concept of public morals exception of GATT Article XX(a), and that in view of protecting this moral concern, the enactment of trade restrictions may be necessary. The EC – Seal Products case is a remarkable milestone in the vindication of the right of sovereign governments to enact trade measures for protecting animals. But the debate around animal welfare measures and the WTO framework is far from being closed. Some old questions remain open (e.g. whether a product originating from a higher-welfare system can be deemed to be like a product originated from a lower-welfare system), and other new ones are emerging.

This paper claimed that the developments around the elaboration of animal welfare recommendations by the World Organisation for Animal Health (OIE) need to gain further attention from international law scholars. There are many potential controversial issues related to the OIE animal welfare recommendation from the perspective of a WTO dispute that are worth exploring in a post-EC – Seal Products case scholarship. This chapter has outlined several questions that could be picked up by future research, going from a historical perspective of the evolving role and characteristics of the OIE to an analytical approach to the problem of assessing compliance with OIE animal welfare recommendations in the context of a trade dispute. Seeking answers to these and other research questions will significantly contribute to the progress of the international governance of animal welfare and the stability and predictability of the multilateral trade system.

Bibliography

- SPS Agreement (1994) Agreement on the Applicationapplication of Sanitarysanitary and Phytosanitaryphytosanitary Measuresmeasures. Marrakesh Agreement Establishing the World Trade Organization, Annex 1A.
- TBT Agreement (1994) Agreement on Technicaltechnical Barriersbarriers to Tradetrade. Marrakesh Agreement Establishing the World Trade Organization, Annex 1A.

Notes

[1] European Communities – Measures Prohibiting the Importation and Marketing of Seal Products (WT/DS400 and WT/DS401).

[2] The abbreviation OIE stands for Office International des Epizooties. In May 2003, the Office became the World Organisation for Animal Health but kept its historical abbreviation.

[3] The GATT Agreement was signed in 1947 and for years served as a provisional arrangement for regulating the international trade of goods. The provisions of the GATT 1947 were incorporated into the WTO Agreement in Annex 1A. This is known as GATT Agreement, 1994.

[4] The concept of the Five Freedoms resulted from a scientific assessment on the conditions of animals kept under intensive farming systems in the UK, The Netherlands and Denmark carried out in 1965. The Brambell Report, as it became known, consists of a document outlining the outcome of the assessment. It stated that animal protection laws should ensure that an animal under an intensive livestock husbandry system had 'sufficient freedom of movement to be able, without difficulty, to turn round, groom itself, get up, lie down and stretch its limbs' (Brambell, 1965, p. 17). Following the publication of this report, the British government installed a governmental body that had the task of giving advice on the revision of current statutory provisions to safeguard the welfare of animals. Building upon the conclusions of the Brambell Report, the Farm Animal Welfare Advisory Committee recommended in 1967 that animals should be free from: (i) hunger and thirst; (ii) discomfort; (iii) pain; (iv) fear and distress; and (v) constraints to express normal behaviour. These five freedoms became a timeless concept that provides a basis for advancing animal welfare science and legislation.

[5] A reference to animal welfare is found in a draft version of the agreement concerning sanitary and phytosanitary measures (SPS Agreement). The reference appears in the text of Annex A.1, where SPS measures were defined. The draft reads as follows: 'measures for the protection of animal welfare and of the environment, as well as of consumer interests and concerns' (for more information see Maciel, 2015, p. 106).

[6] The document G/AG/NG/W/19 is available at: https://docs.wto.org/dol2fe/Pages/FE_Search/FE_S_S009-DP.aspx?language=E&CatalogueIdList=12129&CurrentCatalogueIdIndex=0&FullTextHash=1&HasEnglishRecord=True&HasFrenchRecord=True&HasSpanishRecord=True (accessed 24 November 2018).

[7] In 1990, two trade disputes concerning a measure adopted by the USA to restrict the import of tuna caught in a manner that caused incidental injuries and killings of dolphins have been deemed inconsistent with WTO obligations. Both the Panel and the AB applied a very restricted interpretation of the term 'like product' that supports the principle of non-discrimination of the GATT Agreement. The panels were of the opinion that two physically identical products were to be considered as 'like products' despite differences in their production process methods (PPMs). The USA tried to justify its measure by invoking one of the exceptions listed in GATT Article XX relating to the conservation of exhaustible natural resources. But the Panel found no grounds to uphold a measure aimed at protecting dolphins that were outside the jurisdiction of the importing country.

[8] The regulatory chilling effect is a term used to describe a Member State's decision not to regulate a particular area of environment or social concern due to fears that such a regulation may contravene WTO rules (see Fitzgerald, 2011).

[9] For instance, the dispute European Communities – Measures Affecting Asbestos and Asbestos Containing Products, WT/DS135/R and Add.1, adopted 5 April 2001, as modified by Appellate Body Report WT/DS135/AB/R. The decision of the AB in this case indicated that when establishing if products are 'like', consumer preferences and health effects are elements to be considered.

[10] For more recent literature on this, see Archibald (2008); Sharpless (2008); Cook and Bowles (2010); Howse and Langille (2012).

[11] China – Measures Affecting Trading Rights and Distribution Services for Certain Publications and Audiovisual Entertainment Products, WT/DS363/AB/R, Appellate Body Report adopted 19 January 2010.

[12] United States — Measures Affecting the Cross-Border Supply of Gambling and Betting Services. WT/DS285/ARB, circulated December 21, 2007.

[13] Panel Report, China – Publications and Audiovisual Products, para. 7.759.

[14] The regulation covered products, either processed or unprocessed, deriving or obtained from seals, including meat, oil, blubber, organs, raw or tanned fur skins and articles made from skins and oil. This included articles such as clothing and omega 3 capsules that are made from fur skins and oil from seals.

[15] Panel Report, para 7.505.

[16] Appellate Body Report, para 2.30.

[17] Panel Report, para 7.504.

[18] Appellate Body Report, para 5.338.

[19] For an overview of why those two exceptions have been found to place an arbitrary discrimination, see the EU summary of the case at: http://trade.ec.europa.eu/wtodispute/.

[20] http://www.pig-world.co.uk/news/highlighting-the-differences-how-uk-welfare-standards-compare-with-our-competitors.html

[21] Panel Report, para 7.494.

[22] Panel Report, para 7.494. The Panel also said that 'the principle of minimizing animal pain and suffering gives rise to uncertainty regarding what should be considered an acceptable level of such suffering. Some sources have provided recommendations of humane killing to accommodate the practical demands of seal hunting, thus tolerating risks to welfare that are rejected by others. One notable example in this regard pertains to delays in the killing process, and there has been explicit acknowledgment by some experts of subjectivity and divergence in what is to be considered an acceptable lapse of time between killing steps.'

[23] Panel Report, para 7.495.

[24] Panel Report, para 7.496.

[25] Panel Report, para 7.499.

[26] Panel Report, para 7.496.

[27] As of November 2018, the OIE had a membership of 182 countries.

[28] The complete text of the recommendations is made available at the OIE webpage. There is one animal health code for terrestrial animals (TAHC) and one for aquatic animals (AAHC).

[29] The recommendations draw upon the expertise of internationally renowned specialists to prepare draft texts for new articles of the Terrestrial Code or to revise existing articles in the light of advances in veterinary science. The drafts are then examined by the OIE Scientific Committee and revised by the delegates of member countries, and finally submitted for the deliberation of OIE World Assembly of Delegates. A recommendation is added/revised in the TAHC and AAHC only with the consensus of all OIE members (currently 182 countries).

[30] SPS Agreement – Article 2 (basic rights and obligations).

[31] See, for instance, de dispute settlement 475 – Russian Federation – Measures on the Importation of Live Pigs, Pork and Other Pig Products from the European Union – Recourse to Article 21.5 of the DSU by the European Union – Communication from the Panel.

[32] European Communities – Measures Affecting the Approval and Marketing of Biotech Products. WT/DS291. Panel Report circulated 29 September 2006.

[33] Available at: https://www.iso.org/standard/64749.html (accessed 2 December 2018).

[34] In December 2016, the OIE TAHC was contained in the section of animal welfare Chapters 7.1–7.11. At the time of elaboration of this manuscript the OIE TAHC has 14 sections on its Chapter 7.

References

Archibald, C.J. (2008) Forbidden by the WTO? discrimination against a product when its creation causes harm to the environment or animal welfare. *Natural Resources Journal* 48(1), 15–51.

Brambell, F.W.R. (1965) *Report of the Technical Committee to Enquire into the Welfare of Animals Kept Under Intensive Livestock Husbandry Systems*, Cmnd. 2836 (Great Britain. Parliament). HM Stationery Office, pp. 1–84.

CIWF (2002) WTO: the greatest threat facing animal protection today. Available at: www.ciwf. org.uk (accessed 15 November 2018).

Cook, K. and Bowles, D. (2010) Growing pains: the developing relationship of animal welfare standards and the world trade rules. *Review of European Community & International Environmental Law* 19(2), 227–238. DOI: 10.1111/j.1467-9388.2010.00679.x.

Driver, A. (2017) Highlighting the differences – how UK welfare standards compare with our competitors. The voice of the British pig industry. Available at: www.pig-world.co.uk/ news/ highlighting-the-differences-how-uk-welfare-standards-compare-with-our-competitors.html (accessed 15 November 2018).

Du, M. (2016) Treatment no less favourable and the future of national treatment obligation in GATT article III:4 after *EC–Seal Products*. *World Trade Review* 15(1), 139–163. DOI: 10.1017/ S1474745615000245.

Fitzgerald, P.L. (2011) 'Morality' may not be enough to justify the EU seal products ban: animal welfare meets international trade law. *Journal of International Wildlife Law & Policy* 14(2), 85–136.

GATT Agreement (1994) General Agreement on Tariffs and Trade. Marrakesh Agreement Establishing the World Trade Organization. Annex 1A.

Howse, R., Langille, J. and Sykes, K. (2014) *Sealing the Deal: The WTO's Appellate Body Report in EC – Seal Products* 18(12). Available at: https://www.asil.org/print/1089 (accessed 15 November 2018).

Howse, R. and Langille, J. (2012) Permitting pluralism: the seal products dispute and why the WTO should accept trade restrictions justified by non instrumental moral values. *Yale Journal International Law* 37(2).

Mavroidis, P.C. (2015) Sealed with a doubt: EU, seals, and the WTO. *European Journal of Risk Regulation* 6(3). Available at: https://doi.org/10.1017/S1867299X00004839 (accessed 13 September 2018).

Möllenhoff, J. (2015) Framing the 'public morals' exception after EC – seal products with insights from the ECTHR and the GATT national security exception. The Graduate Institute of International and Development Studies, Centre for Trade and Economic Integration. *CTEI Working Paper*, 2015–2017.

Jenkins, L. and Stumberg, R. (2001) Animal protection in a world dominated by the World Trade Organization. In: Salem, D.J. and Rowan, A.N. (eds) *The State of the Animals*. Humane Society Press, Washington, DC.

Maciel, C.T. (2015) Public morals in private hands? A study into the evolving path of farm animal welfare governance. PhD thesis. Wageningen University, Wageningen, The Netherlands.

Offor, I.H. and Walter, J. (2017) GATT article XX(a) permits otherwise trade-restrictive animal welfare measures. *Global Trade and Customs Journal* 12(4), 158–166.

OIE (2019) *Terrestrial Animal Health Code: General Provisions*. Vol. 1. World Organisation for Animal Health.

Sttuhent, P. (2017) The use of public morals exception for animal welfare in the WTO: EC-Seal products. *Research Gate*. DOI: Available%20at: https://www.researchgate.net/publication/323415051_The_Use_of_Public_Morals_Exception_for_Animal_Welfare_in_the_WTO_EC-Seal_Products (accessed 16 February 2020)..

Cottier, T. (2018) The implications of EC – seal products for the protection of core labour standards in WTO law. *Labour Standards in International Economic Law.*

Sharpless, I. (2008) Farm animal welfare and WTO law: assessing the legality of policy measures. Master of Arts in Law and Diplomacy Thesis. The Fletcher School, USA.

Stevenson, P. (2000) The world trade organisation: the need to end its detrimental impact on animal welfare. Doc PS/MJ/ART8502. Available at: http://trade.ec.europa.eu /doclib/docs/2005/april/tradoc_122236.pdf (accessed 22 September 2011).

Stevenson, P. (2002) The world trade organisation rules: a legal analysis of their adverse impact on animal welfare. Available at: https://heinonline.org/HOL/LandingPage?handle=hein.journals/anim8&div=8&id=&page= (accessed 15 November 2018).

Stevenson, P. (2009) European and international legislation: a way forward for the protection of farm animals? In: Sankoff, P. and White, S. (eds) *Animal Law in Australasia Federation Press Sydney*, pp. 307–332. Available at: http://trade.ec.europa.eu/doclib/docs/2013/february/tradoc_150460.pdf

Thomas, E.M. (2007) Playing chicken at the WTO: defending an animal welfare-based trade restriction under the GATTs moral exception. Boston College Environmental Affairs Law Review 34. Available at: https://lawdigitalcommons.bc.edu/ealr/vol34/iss3/8/ (accessed 15 November 2018).

Veissier, I., Butterworth, A., Bock, B. and Roe, E. (2008) European approaches to ensure good animal welfare. *Applied Animal Behaviour Science* 113(4), 279–297. DOI: 10.1016/j.applanim.2008.01.008.

White, S. (2013) Into the void: international law and protection of animal welfare. Available at: https://doi.org/10.1111/1758-5899.12076 (accessed 13 September 2018).

WTO (2015) European communities – measures prohibiting the importation and marketing of seal products. addendum WT/DS400/16/Add.7WT/ DS401/17/ Add.7. 16 October 2015. Available at: https://www.wto.org/english/ tratop_e/dispu_e/cases_e/ds400_e.htm (accessed 15 November 2018).

9 Farm Animal Welfare: The Future

ALISTAIR STOTT,[1] AND BOUDA VOSOUGH AHMADI[2*]

[1]Retired professor of animal-health and welfare economics
[2]The European Commission for the Control of Foot-and-Mouth Disease (EuFMD) of the Food and Agriculture Organization of the United Nations (FAO)

Summary

Science can help us understand what animals want and economics can provide the understanding of human motivation needed to deliver such wants. In our view, what needs further development in future is for economics and information/communication science and technology to channel awareness into appropriate action. This chapter elaborates on this idea by providing some illustrative examples. Focusing on animal health and welfare, it argues that there is much scope for improvement in profit and welfare on commercial farms simply by adopting the best disease management approach available. We also emphasize the importance of systems modelling and operations research (OR) in the future to ensure that animal welfare taps into the growing opportunities that developments in these methods are likely to bring. The chapter also argues that OR can provide a bridge between animal welfare science, economics and business to deliver improvements in animal welfare through food markets. The importance of big data and precision livestock farming in livestock production/reproduction, animal health and welfare, and the environmental impact of livestock production are also discussed. New genetic approaches to optimize livestock resilience and efficiency are highlighted. We argue that tackling difficult problems, such as sustainability (that encompasses animal welfare alongside environment and climate change), efficiency and resilience in farm animal production systems, is and will remain a vital focus of research in the agri-food sector. Research methods and governance still need to change to properly reflect this. It is

*Corresponding author: Bouda.ahmadi@fao.org

envisaged that animal welfare will be affected by these developments and should, wherever appropriate, be explicitly considered.

9.1   Introduction

It is important in a text such as this to establish clear, complementary working definitions of animal welfare and of economics. Buller *et al.* (2018) maintain that much disagreement about animal welfare stems from mixing up ethical concerns about how we should treat animals with scientific questions about their welfare *per se*. However, they concede that the scientific concept of animal welfare is still developing and arose out of ethical concerns about how we treat animals. Further details about this ongoing process are given by Lawrence and Vigors in Chapter 1 of this volume. Neeteson *et al.*, in Chapter 6, cite Dawkins (1980) for a definition of animal welfare as 'animals that are healthy and have what they want'. They warn against determining animal welfare in terms of anthropomorphism and human perspectives on freedom to express 'natural' behaviours, citing Dawkins (2012).

Definitions of economics are equally difficult. Robbins (1935), in an essay on the nature and significance of economics, stated that we all talk about the same things, but we have not yet agreed what it is we are talking about. He goes on to propose a definition of economics as the science that studies human behaviour as a relationship between ends, and scarce means, which have alternative uses. This now widely quoted definition highlights the importance of scarcity and choice in economics (Ritson, 1978). Also widely quoted is the definition of economics by Adam Smith (1776) as the study of people's rational pursuit of self-interest. As he put it: 'It is not from the benevolence of the butcher, the brewer, or the baker that we expect our dinner, but from their regard to their own interest.' This adds an important additional element to the Robbins definition by the inclusion of motivation. Human motivation drives improvements in animal welfare and is implicit in all other chapters of this book. It follows that if science can help us understand what animals want and economics can provide the understanding of human motivation needed to deliver such wants in a competitive environment, then we have complementary working definitions of the two disciplines.

Given the above joint definition, it is important for the future of animal welfare that science and economics work closely together. This should be inclusive of other disciplines such as psychology, sociology and information/communication science and technology (ICT). In this way, human decision making that affects animal welfare may be based on what delivers evidenced animal benefits that provide the psychological and/or other rewards people expect in return. This should fuel further positive actions. Economists such as McInerney (1993) have argued that animal welfare benefits only accrue insofar as they meet human needs. Lusk and Norwood (2011) contrast this position with the utilitarian argument that animal welfare should be considered for inclusion in economic assessments of overall human welfare on the grounds that animals suffer. From both standpoints, it is human perceptions of animal welfare that drive the 'self-interested' human behavioural change that may or may not deliver the improved welfare that the animal would recognize.

The potential for change in the future rests on science and other disciplines providing and delivering the evidence that economics can help to turn into the necessary behavioural change. This is a development of the arguments that Lawrence and Vigors put forward as one of the roles of science in animal welfare in Chapter 1. They base their comments on Serpell (2004) under the title 'science as a cultural moderator'. Their point is that science can raise awareness of and positive interest in animals and their ecology. Since Serpell, social media has increased the opportunities for this approach. Similar recent developments in environmental awareness (e.g. the 'Blue Planet' TV series, BBC, 2019) show what can be achieved. What needs further development in future is for economics and ICT to channel awareness into appropriate action. This chapter is based on this idea.

9.2   Animal Health

Animal health is clearly an important aspect of animal welfare. In farm animals, the financial benefits to farmers are often perceived as a sufficient motivation to actions that will deliver health and welfare improvements (Buller *et al.*, 2018). However, high incidence of costly endemic diseases such as mastitis and lameness persist (Hogeveen *et al.*, 2011; Langford and Stott, 2012) and the risk of exotic disease incursions must be guarded against by both farmers and policy makers. For example, Barratt *et al.* (2019) estimated a range of losses from a future simulated foot-and-mouth disease (FMD) outbreak in Scotland at £400–950 million. The National Audit Office (2002) put the cost of the notorious UK outbreak of FMD in 2001 at £8 billion. The scale of the human and animal suffering that accompanied this event is not directly captured in the financial figures but must be proportional to them. Human suffering via animal disease, for example through zoonotic infections and links to antimicrobial resistance (O'Neill, 2014), are likely to increase the importance of this aspect of animal welfare still further in the future.

Economics that simply ascribes costs to animal disease is of little, if any, value (Mclnerney, 1996). What matters in an environment where there is competition for scarce resources is to establish the best control choices based on likely cost-effectiveness. McInerney *et al.* (1992) proposed a simple theoretical framework to do this. Yalcin *et al.* (1999) applied this framework in practice, as shown in Fig. 9.1.

Each cross in Fig. 9.1 represents the average total financial costs (losses due to disease plus control expenditure) due to subclinical mastitis for a cohort of Scottish dairy herds using the same set of control strategies. There were 750 herds in the sample and 28 distinct control cohorts. Clearly, cohorts with greatest control expenditure tend to have the smallest losses. However, the cohorts with the smallest losses at any given control expenditure are most cost-effective and are highlighted by the solid line (loss-expenditure frontier).

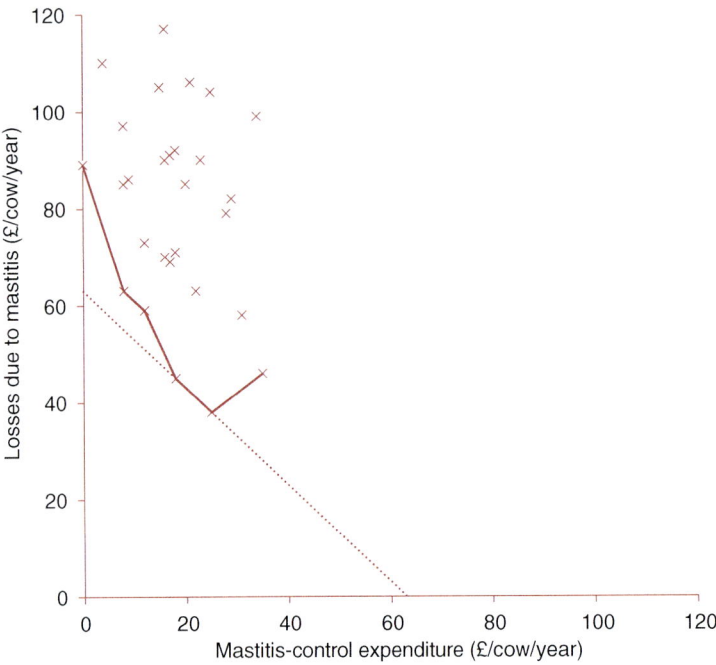

Fig. 9.1. Loss-expenditure frontier (LEF) for subclinical mastitis in Scottish dairy herds with high somatic cell counts. (From Yalcin *et al.*, 1999)

The dotted line joins all combinations of loss and control monies that sum to the same total cost. As this second line is tangential to the first it identifies the cohort with the minimum total cost of mastitis, i.e. £66/cow/year. The average total cost in the sample was £100/cow/year. This means that the sample could avoid excess costs of £34/cow/year on average by switching to the most cost-effective control strategy. In this example, the best control strategy occurs at the lowest level of losses (greatest welfare), i.e. a 'win–win' situation for farm profits and for animal welfare. This will not always be the case, as explained in Chapter 5 of this book by Niemi. The example serves to illustrate how much scope there is for improvements in profit and welfare on commercial farms simply by adopting the best disease-management approach available. It also shows how much variation exists even in these averaged figures, with the worst cohorts incurring almost twice the output losses, and so considerably more disease, and hence welfare challenge, than should be necessary. This example demonstrates the power of economics to address welfare concerns provided that the information, and the motivation to use it, are available. The opportunity is not of course confined to mastitis or even just to disease challenges to welfare.

Economics can help policy makers and advisers establish priorities for investment in research, extension or legislation to bring down disease prevalence either by incentives to farmers to adopt evidence-based improvements in existing disease control and prevention and/or to provide new technologies that might shift the loss-expenditure frontier downwards. So far, use of these objective interdisciplinary approaches in these ways has been limited. Developments of ICT could possibly remedy this in future.

To provide bespoke improved disease and welfare decision support for the individual farm/farmer requires an alternative framework to the population-based loss-expenditure frontier method. To address this problem, Stott and Gunn (2008) adapted the benefit function methodology of Tisdell (1995) and applied it to simulated bovine viral diarrhoea virus (BVDV) infection in a typical Scottish beef suckler herd, as shown in Fig. 9.2. This provides an applied example of the production economics theory described by Niemi in Chapter 5. This time, the hatched line from the origin at 45 degrees represents the break-even level of investment. The vertical lines under each scenario represent the profit-maximizing level of investment furthest away from break-even where marginal cost of biosecurity equals marginal benefit. In each scenario, profit-maximizing levels of investment fall short of the benefit-maximizing position. In this example, there is no win–win between profit and welfare. It is likely that animal welfare will benefit from improvements in biosecurity even beyond the financial benefit-maximizing point, as small marginal reductions in the risk of BVDV incursion continue at ever-increasing marginal cost. This win–lose example seems more plausible in most scenarios than the previous win–win example. Diminishing returns apply, and so the cost of the highest levels of welfare are likely to be prohibitive unless a market price premium, government subsidy or third-party

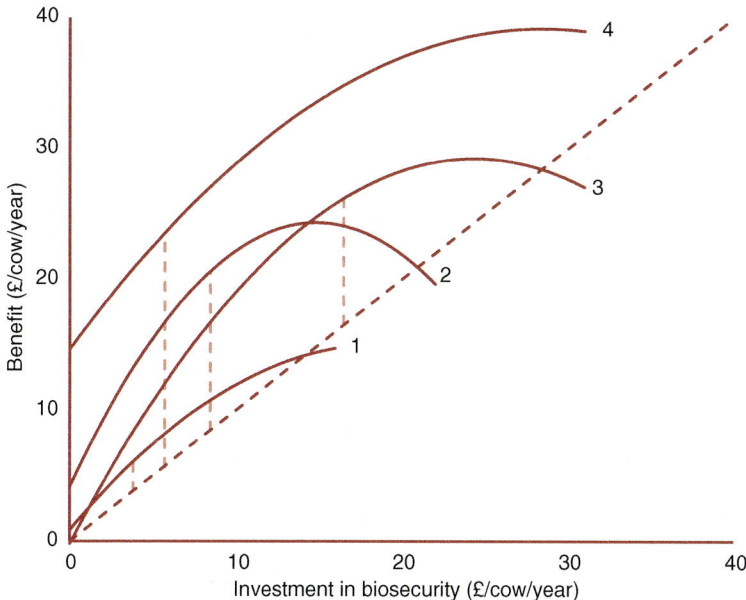

Fig. 9.2. Benefit function for increasing investment in biosecurity against BVDV incursion under four alternative scenarios in a simulated 50-cow suckler herd. (From Stott and Gunn (2017))

contribution can be found. At least frameworks such as this can help to establish what a fair cost-sharing agreement might be to ensure farms remain competitive, while providing higher levels of welfare input than may otherwise be carried by the free market alone. Where externalities like animal welfare and environmental pollution, etc. exist (see Hubbard *et al.*, Chapter 2), the issue of reducing transaction costs so that the market may better facilitate payments by 'polluters' to those affected, so as to reduce pollution, is much debated (e.g. Ekboir, 1999). This may be an important issue for animal welfare economics in the future.

An interesting adjunct to Fig. 9.2 is that scenarios 3 and 4 assume herd disease status is known, i.e. the herds are in a health scheme, while scenarios 2 and 4 assume vaccination against BVDV. Greatest net benefit ('profit') from investment in biosecurity by far accrues to scenario 4 where both vaccination and health scheme membership are in place. Insofar as these aspects confer improved animal welfare, there is a win–win. However, both substitute for some investment in biosecurity in scenario 4 but not to the low level justified where health status is unknown and no vaccination takes place (scenario 1). Also note that in scenario 4, much greater than optimal levels of biosecurity investment would bring very small reductions in net benefit, i.e. such farms could be easily persuaded to make further investment for public and/or private benefit beyond the private BVDV benefits included here. Being able to tease out these interactions and demonstrate a considerable financial benefit (motivation) to farmers for health scheme membership, vaccination and biosecurity is a clear animal welfare benefit for this type of analysis.

9.3   Systems Modelling

The BVDV example above was made possible by the use of a simulation model predicting the spread of the virus over time as it interacts with the normal cycle of events that represent the farming system concerned (suckler beef (cow-calf) herds) (see Stott *et al.*, 2003, and Gunn *et al.*, 2004, for further details). In the case of animal disease, it is often unethical to gather necessary data, e.g. from unchecked epidemics, so simulation models are the only way to progress the search for strategies that improve animal health and welfare while estimating the financial incentives involved. However, the case of animal health illustrates the general problem that data to establish trade-offs (biological interactions) between animal welfare and other biological functions that impact productivity and economics are often lacking (Lawrence and Vigors, Chapter 1). Systems modelling is a way forward. Although such models will often be a gross simplification of the processes involved, they may thereby reduce the dependence on data that describe these processes, and (with appropriate testing of model validity) still more objectively and reliably assess decision choices that affect animal and human welfare than would otherwise be possible. Recent developments in ICT will reduce these problems and increase the scope and power of model applications (Janssen *et al.*, 2017). Interdisciplinary collaboration is also a challenge in systems modelling that can be addressed (Kragt *et al.*, 2016). It will therefore be important in future to ensure that animal welfare taps into the growing opportunities that developments in systems modelling are likely to bring. The implications for animal welfare (intended or not) from greater use of systems modelling in general agricultural and food decision support must also be monitored.

It is possible to lead systems models in animal welfare from the economic (human behaviour/scarce resource choices) standpoint rather than from the biological processes concerned. The latter was used in the above animal health examples. Examples of the former approach cited elsewhere in this volume include Ahmadi *et al.* (2011) where linear programming (LP) was used to optimize the financial performance of alternative farrowing systems subject to specific welfare and managerial constraints. This allowed comparison of the alternative systems on a common basis and established the minimum trade-off and maximum synergies between welfare attributes and financial performance within and between systems. Barnes *et al.* (2011) (also cited elsewhere in this book) used data envelopment analysis (DEA) to examine the technical efficiency of a sample of dairy farms categorized by lameness as a proxy measure of welfare. DEA places the best farms in the sample on an efficiency frontier with which all the other farms can then be compared. This is not dissimilar to the LEF approach described above and illustrated in Fig. 9.1. Barnes *et al.* (2011) established that low-lameness farms had the highest technical efficiency but had low efficiencies in terms of labour and stocking density. This highlighted the importance of approaches to animal welfare assessment that capture the whole farm system and not just a particular aspect. On farms where labour and land are particularly scarce, lameness is likely to be more of a challenge if the business is to compete successfully with farms where such resources are less constrained. It follows that improving welfare among heterogeneous farm businesses requires a bespoke approach that systems models can help to support. The example also highlights the difficulties of using simple metrics of animal welfare or even a farm assurance scheme based on a basket of such metrics (FAWC, 2005) in food marketing. If these schemes are to develop in future in ways that best improve animal welfare and food security together through food markets, then closer integration between science and economics will be required. This will ensure that assurance schemes reflect important differences between farm businesses and hence variation in their priorities for action.

Systems models that primarily address human decision making are often drawn from management sciences (sometimes referred to as operations research or OR). This discipline follows the scientific method, drawing from various other disciplines including mathematics, statistics and engineering (Taylor, 2016). Their key feature is a focus on solutions to practical business problems, i.e. they provide applied support to economics as defined in the above introduction. So far, few applications have been directly applied to animal welfare challenges. An exception is Stott *et al.* (2012) who used OR both to measure animal welfare and to address the problem of feeding sheep throughout the year on extensive hill-farming systems in Great

Britain. The OR-based welfare measurement (Service Quality Modelling (SQM) (Parasuaman *et al.*, 1988)) provided a complementary and comparable outcome to more usual ('needs'-based (Keeling and Veissier, 2005)) scientific approaches. The feeding element of the modelling was based on an LP application, which optimized farm resource use (land, labour, grazing and forage availability) month-by-month over the year to deliver the maximum possible sheep enterprise gross margin (GM) for a given farm inventory. This approach provided an objective method to establish GM, based on readily available farm management data, a model of grass growth (Armstrong *et al.*, 1997) and the well-established metabolizable energy (ME) system for livestock feed rationing (AFRC, 1993). In this way, comparable gross margins were established for 20 hill sheep farms. The LP established that in extensive sheep-farming systems, greater productivity was associated with higher overall welfare scores. This suggested that such farms lie to the left of the theoretical curve proposed by McInerney (1994), reproduced in Fig. 9.3. In this way, they contrast with most other more intensive livestock systems assumed to lie in a trade-off position 'C'. Unfortunately, increased productivity and welfare score were not associated with increased gross margin in the model of Stott *et al.* (2012). Greatest financial return was found in flock expansion with reduced labour per ewe and no increase in productivity, i.e. a win–lose situation. The example serves to demonstrate the 'what if?' capabilities of such systems models that can aid decision support both at farm and policy level. OR is well established in food supply chain businesses (He *et al.*, 2018) and so is most likely to provide a future bridge between animal welfare science, economics and business to deliver improvements in animal welfare through food markets. With such models, it will be easier to address the complex multi-attribute nature of animal welfare and the need for marginal improvements to be balanced by the cost of other products and attributes forgone. Hubbard *et al.* have covered these points in Chapter 2 of this volume. They highlight the dearth of research in this area and the associated

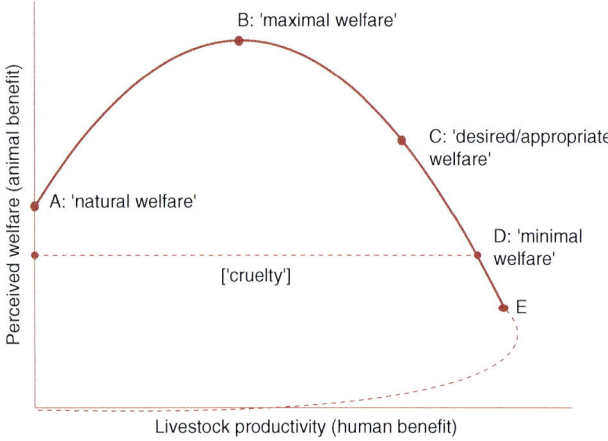

Fig. 9.3. Trade-offs between human and farm animal welfare proposed by McInerney (1994).

problem of turning consumers' willingness to pay for animal welfare improvements into the necessary purchase decisions.

9.4   Big Data

The previous section highlighted the advantages of systems models that can overcome lack of data in order to support better decision making throughout the food supply chain leading to improved animal welfare. Developments in big data promise to overcome rather than circumvent the problem (Kamilaris *et al.*, 2017). The future opportunities in this area for animal welfare are too broad, diverse and nascent to list here. Instead, we raise some illustrative examples.

Kehlbacher *et al.* (2012) established that consumers' willingness to pay (WTP) for meat produced to different welfare standards is sufficiently differentiated to make a labelling scheme in the form of a certified rating system appear feasible. Big data offers the opportunity to realize the full potential of such a system by addressing the problems of imperfect information, competing demands and cognitive dissonance faced by consumers (see Bennett, Chapter 4). It is conceivable that in future, automatic data capture from mobile phones, social media, online purchasing history, data from the 'internet of things' (smart fridges and freezers), etc. may be brought together with data from the food marketing chain through machine learning to produce ethical purchasing recommendations for individuals or groups of consumers. These systems may not be confined to purchasing decisions and could encompass, for example, charitable giving, shareholding in ethical farming/food companies and citizens' contributions to public policy making.

Wemelsfelder *et al.* (2001) established that qualitative behavioural assessment (QBA) of farm animals by human observers is valid as an integrative welfare assessment tool. Using automated data capture to gather and pool data from video cameras, feeders, drinkers, the animals' environment and physiology, etc. it may be possible to calibrate these data against human observations of QBA and other welfare indicators. The system could then form a closed and/or open feedback loop to ensure that animal welfare is safeguarded at all times by timely interventions to adjust the environment and provide inputs as necessary. In this way, it may be possible, for example, to detect subclinical disease incidence in an individual or herd and stamp the problem out before welfare is compromised.

Precision livestock farming provides a management system based on automatic real-time monitoring and control of animal production/reproduction, health, welfare and the environmental impact of livestock production (Berckmans, 2014). For example, Lambe *et al.* (2015) describe the potential opportunities associated with electronic identification of sheep, which was made compulsory in the UK in 2010. Benefits for animal welfare include more accurate record keeping, less stressful handling and individual monitoring of animal performance and health. The latter, for example, facilitates targeted use of anthelmintics based on live weight change, thus reducing drug use and increasing its effectiveness (McBean *et al.*, 2016).

9.5   Animal Breeding

Developments of new technology in animal breeding will make it increasingly feasible to aim for genetic improvement of productivity, product quality and animal adaptability simultaneously (see Chapter 6). Broom, in Chapter 7, highlights EU Horizon 2020 research programmes 'breeding livestock for resilience and efficiency' (SFS-15-2016-2017) that will address animal welfare challenges through the use of animal breeding and genetics. The GenTORE project (http://www.gentore.eu/) is an example. The objective of GenTORE is to develop new genome-enabled selection and management tools to optimize cattle resilience and efficiency in widely varying environments (Friggens *et al.*, 2017). Optimization and the criteria by which it is judged will be a key challenge. There is a balance to be struck between resilience and efficiency, which will vary between environments. Balances also exist between selection traits, which, as Neeteson *et al.* point out in Chapter 6, are often negatively genetically correlated. These traits will have trade-offs between productivity, profit, environment and welfare (see Fig. 9.3) that will vary over time. Animal breeding must also be integrated with other tools at farm level to deliver efficiency and resilience. Beyond these farm-/animal-based projects, there is also a need to balance contributions from livestock agriculture to sustainable food security alongside other initiatives and their wider socio-economic impacts. Although projects such as GenTORE include socio-economics, there is a general tendency to drive research and development via new technology and scientific innovation rather than taking a more interdisciplinary, problem-centred approach (Kragt *et al.*, 2016) where economics, as defined in the above introduction, is central. Tackling difficult problems such as sustainability, efficiency and resilience is and will remain a vital focus of research in the agri-food sector. Research methods and governance will need to change to reflect this. Animal welfare will be affected by these developments and should, wherever appropriate, be explicitly considered.

Economics has been used to assess the value of alternative animal breeding strategies. For example, Fig. 9.4 shows responses to alternative expanded selection indices used in the UK dairy cattle breeding programme. As the index expands across the *x*-axis, the financial response improves. This is because the economic values used in the index were derived using an OR systems model to maximize the net benefit to producers from a marginal improvement in lifespan, mastitis incidence and lameness from a base index (PIN) confined to milk production values. In all cases, mastitis and lameness responses are positive, i.e. they get worse with use of the farm profit-oriented selection index. This is because the financial benefits of increasing milk yield by genetic selection outweigh the losses from mastitis and lameness caused by the negative genetic correlations involved.

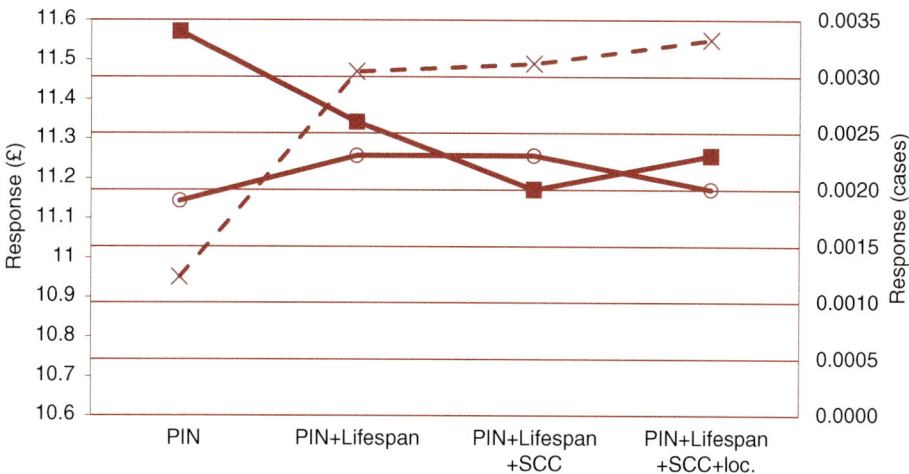

Fig. 9.4. Predicted responses to alternative selection indices in the UK dairy cattle sector based on Stott *et al.*, 2005b. The hatched line shows the financial response (left axis). The solid line with solid markers tracks the mastitis response (right axis). The remaining line tracks lameness response (right axis).

However, inclusion of lifespan and SCC (somatic cell count, a measure of mastitis) progressively reduces the undesirable mastitis response (win–win). Inclusion of a lameness trait in the final index restores the rate of decline in lameness to its level with PIN alone but at the expense of a small deterioration in mastitis response. The example illustrates the complexity of the trade-offs involved in animal breeding between welfare-related traits and productivity. The interdisciplinary work at least makes these trade-offs explicit, providing an opportunity for policy intervention. Since this work was done, goal traits have been expanded still further, most notably to include environment-/climate change-related improvements. The problem of trade-offs remains, as does the issue that new concerns are often public goods like animal welfare and the environment, yet are usually paid for by industry for private benefit (see Chapter 2).

9.6   The Environment

The Foresight Report (Foresight, UK, 2011) highlighted the 'perfect storm' where expanding global population, increased per capita demand (including for livestock products), climate change (possible reduced food supplies) and dwindling non-renewable resources come together to build an unprecedented crisis. The response so far in agriculture has majored on increasing productivity to reduce greenhouse gas (GHG) emissions per kg of product (Llonch *et al.*, 2017), as livestock are thought to directly contribute about 11% of anthropogenic GHG (Gerber *et al.*, 2013). This has serious implications for animal welfare (Fig. 9.3), which will vary between livestock systems as shown above. The safeguard suggested by Foresight was 'sustainable inten-sification' (SI). But does this encompass animal welfare or is it confined to maintaining environmental services for future generations? The answer to this question will depend on how SI is achieved in future. As demonstrated above, interdisciplinary research, aided by ICT, provides a basis for the de-velopment of analytical frameworks to ensure that SI properly addresses all three pillars of agricultural sustainability (food security, environment and socio-economic, the latter including animal welfare (Barnes and Thomson, 2014)).

Llonch *et al.* (2017) review current strategies to mitigate GHG in live-stock systems with particular reference to animal welfare. They maintain that some strategies will increase welfare alongside productivity (e.g. im-proved animal health and nutrition). Strategies that constrain the animals' environment, and so impinge on their freedom to exhibit normal behaviours, may compromise animal welfare. However, there is more to sustainable live-stock production than GHG emissions. As Eisler *et al.* (2014) point out, ani-mal products offer particular human dietary benefits. Animals can convert food waste, by-products and forage into human food, clothing, fuel, fertiliser and power. Sustainably managed grazing can increase biodiversity, maintain ecosystem services and improve carbon capture by plants and soil. These authors also point out the socio-cultural importance of livestock as collateral, a mark of social status, the basis for business development opportunities, etc. particularly among the world's poorest peoples.

Given the global importance of livestock to the environment, climate change and socio-economic development (including animal welfare con-cerns), it is clearly vital to ensure that associated decision support frame-works match up. Environmental economics now has widely accepted frameworks such as marginal abatement cost curves (MACC). The MACC ranks GHG abatement strategies by their cost per unit of abatement (verti-cal axis) and volumes abated (horizontal axis). In this way, strategies that have the greatest impact at least cost (or most benefit) are easily identified. For example, Moran *et al.* (2011) found that the equivalent of 5.38 metric tonnes of CO_2 could be abated annually at negative or zero cost in UK agriculture by 2022 (win–win). This represents about 12% of total emis-sions from UK agriculture. Livestock strategies in this category included improved genetics, productivity, nutrition (ionophores) and fertility. The animal welfare implications of these improvements can and should be monitored. Where trade-offs exist, ways must be found to resolve these, such as those illustrated above.

Another widely used and hence influential approach to decision support in this area is Life Cycle Analysis (LCA). This technique examines the envi-ronmental impacts of products across their life cycle, capturing the relative importance of different sectors and interactions between them, and hence the wider consequences of alternative actions. Despite this, Garnett and Godfray (2012) argue that LCA provides only a limited understanding of livestock GHG emissions. A broader framework is required that includes in-direct effects of land use change, the opportunity costs of resources devoted to livestock and the need for livestock products. These issues clearly have parallels in animal welfare and hence opportunities to incorporate animal welfare in such frameworks (Tallentire *et al.*, 2019). For example, Soteriades *et al.* (2016) combine LCA with DEA to examine how dairy farm management affects environmental outcomes at regional level. Given the necessary data, the approach provides the more holistic perspective needed to study farm sustainability and incorporate animal welfare specifically, as was done with DEA by Barnes *et al.* (2011).

9.7   Behavioural Economics

As Hubbard *et al.* in Chapter 2 point out, very few people are simple *homus economicus*, i.e. their rational behaviour highlighted by Smith (1776) is not governed solely by the optimal solution to a problem that might emerge from an OR model, DEA or other data-driven approach. Such models will rarely, if ever, deliver perfect information. Farmers, for example, may therefore be better persuaded of the case for investing in animal health and welfare based on the relative risks involved rather than by a cost-benefit analysis (Stott and Gunn, 2017). Decision makers respond to their peers and social norms, which can be 'nudged' by the provision of appropriate information and the opportunities to examine and debate it in support of a decision (Thaler and Sunstein, 2008). These points have given rise to the discipline of behavioural economics. Given the nature of animal welfare and the human decisions that influence it, from farm to supermarket shelf and all steps in between, it seems likely that behavioural economics will play an important role in future animal welfare policy. A key element of behavioural economics is to understand what drives decisions and from that understanding to create decision support that nudges towards preferred outcomes. Who does the nudging and why is, of course, a key issue. In public animal welfare research, the conceptual framework of libertarian paternalism is usually assumed. This requires policies to protect individual liberty, focus on the benefit of those targeted and be informed by the findings of behavioural economics (Oliver, 2015). Some illustrative examples follow.

Using data from 625 consumers in England, Toma *et al.* (2012) established *a priori* determinants of behaviour that govern purchase of free-range and organic chicken meat. In both cases, attitudes towards animal welfare and access to information on animal welfare issues were significant drivers, among others such as education and price. The nature and strength of the interrelationships between drivers provide information on what nudges would best increase purchase of, and hence support for, higher welfare production systems. Studies of this type when aligned with the science of animal behaviour and welfare may help to ensure that food marketing delivers greater real improvements in animal welfare (see Chapters 3 and 4).

It is particularly important in studies of sustainability in agriculture that decision support frameworks capture the full spectrum of behavioural as well as environmental concerns. For example, Santarossa *et al.* (2004) used principles of natural resource economics to determine the maximum sustainable yield per hectare in the dairy sector. This contrasted with the more usual focus on maximizing gross margin in the short run. The approach recognized that the dairy herd must divert resources (e.g. land) to heifer rearing in order to sustain productive cows and calves. A common response is to farm more intensively. However, a consequence of this may be reduced animal welfare and degradation of land (soil erosion). A balance must therefore be struck between current and future milk and meat production and associated trends in land productivity and value. This balance revolves around the ratio of cows to heifers, which in turn depends on cow fertility (conception rate). The authors therefore focused on establishing the optimal conception rate based on both current profits from milk and meat production and the capital (productive) value of land including a terminal land value. The capital value of land was linked to farming intensity via soil quality, climate, grass growth, dietary energy budget and nitrogen recycling. Optimum conception rate fell from 0.88 to 0.72 when the capital value of land was included, implying less pressure on the cows and hence improved welfare. From the farmer/landowner's point of view, short-term profits are balanced against the long-term value of the land and its final resale value on retirement. The environmental footprint of the system is assessed directly in terms of sustaining land capability rather than just GHG emissions, although comparative carbon footprints could be added to the model.

Figure 9.5 shows the relationship between welfare score and net farm income for a specific hill farm in Great Britain contributing to the study reported by Stott *et al.*, 2005a. Welfare scores were based on the combined responses of sheep farmers to a computer-led survey based on adaptive conjoint analysis (ACA). ACA is a market research tool more often used to explore consumer perceptions of novel products and to help understand purchasing behaviour (Green and Srinivasan, 1990). The graph shows how housing ewes and reducing the number of hill gatherings can improve welfare and profit at lambing time. Scanning pregnant ewes gave a considerable boost in welfare

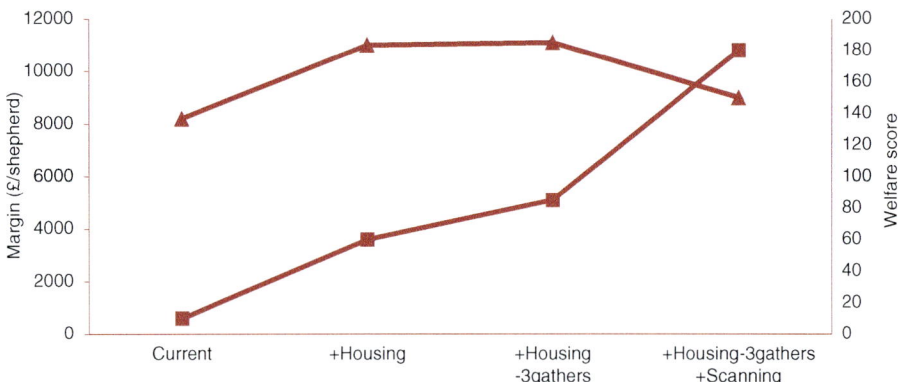

Fig. 9.5. Unpublished welfare development programme for an extensive hill sheep farm contributing to a project reported by Stott *et al.* (2005a).

but was detrimental to farm profits. This and other approaches were used in a participative research project with farmers (Defra, 2005) to explore, share and nudge farmer behaviours towards improved welfare and farm business management.

9.8   International Trade Aspect

Understanding of the need to expand animal protection laws emerged first among European countries, and in recent years it has also been embraced by other countries. Thus the fact that animal welfare was not yet a widely recognized need in the 1990s explains why earlier attempts failed to include a reference to animal welfare in one of the agreements leading up to the creation of the WTO system (see Chapter 8). In this chapter, the authors argue that a thoughtful consideration of what role OIE animal welfare recommendations will eventually play in a trade dispute is still missing. After almost two decades of uncertainty and controversies over the lawfulness of restricting trade on the basis of animal welfare concerns, the ruling in the EC – Seal Product case confirmed that citizen concerns over the welfare of animals falls within the content and scope of the concept of public morals exception of GATT. Therefore, in view of protecting this moral concern, the enactment of trade restrictions may be necessary. The EC – Seal Products case is a remarkable milestone in the vindication of the right of sovereign governments to enact trade measures for protecting animals. However, the debate around animal welfare measures and the WTO framework is far from being closed. Some old questions remain open and other new ones are emerging. Brexit is likely to highlight these issues in the future as the UK, as a non-member, strives to renegotiate trade links with the EU and to establish new trade links across the rest of the world.

Acknowledgements

The authors acknowledge their current and past research funders as well as many colleagues and co-authors who contributed enormously to their research and publications.

References

AFRC (1993) *Energy and Protein Requirements of Ruminants. An Advisory Manual Prepared by the AFRC Technical Committee on Responses to Nutrients.* CAB International, Wallingford, UK.

Ahmadi, B.V., Stott, A.W., Baxter, E.M., Lawrence, A.B. and Edwards, S.A. (2011) Animal welfare and economic optimisation of farrowing systems. *Animal Welfare* 20, 57–67.

Armstrong, H.M., Gordon, I.J., Grant, S.A., Hutchings, N.J., Milne, J.A. *et al.* (1997) A model of the grazing of hill vegetation by sheep in the UK. I. The prediction of vegetation biomass. *The Journal of Applied Ecology* 34(1), 166–185.

Barnes, A.P., Rutherford, K.M.D., Langford, F.M. and Haskell, M.J. (2011) The impact of lameness prevalence on dairy farm level technical efficiency: an adjusted data envelopment analysis approach. *Journal of Dairy Science* 94, 5549–5557.

Barnes, A.P. and Thomson, S.G. (2014) Measuring progress towards sustainable intensification: how far can secondary data go? *Ecological Indicators* 36, 213–220.

Barratt, A.S., Rich, K.M., Eze, J.I., Porphyre, T., Gunn, G.J. *et al.* (2019) Framework for estimating indirect costs in animal health using time series analysis. *Frontiers in Veterinary Science* 6, 190. DOI: 10.3389/fvets.2019.00190.

BBC (2019) BBC One - The Blue Planet. Available at: https://www.bbc.co.uk/programmes/b008044n (accessed 19 June 2019).

Berckmans, D. (2014) Precision livestock farming technologies for welfare management in intensive livestock systems. *Revue Scientifique et Technique de l'OIE* 33(1), 189–196. DOI: 10.20506/rst.33.1.2273.

Buller, H., Blokhuis, H., Jensen, P. and Keeling, L. (2018) Towards farm animal welfare and sustainability. *Animals* 8(6), 81.

Dawkins, M. (1980) *Animal Suffering: The Science of Animal Welfare*. Oxford University Press, Oxford, UK.

Dawkins, M. (2012) What do animals want? A conversation with Marian Stamp Dawkins. The Reality Club. Available at: https://www.edge.org/conversation/marian_stamp_dawkins-what-do-animals-want (accessed 14 October 2018).

Defra (2005) Improving sheep welfare on extensively managed flocks; economics, husbandry & welfare - AW1012. Available at: http://sciencesearch.defra.gov.uk/Default.aspx?Menu=Menu&Module=More&Location=None&Completed=0&ProjectID=10988 (accessed 17 June 2020).

Eisler, M.C., Lee, M.R.F., Tarlton, J.F., Martin, G.B., Beddington, J. *et al.* (2014) Agriculture: steps to sustainable livestock. *Nature* 507(7490), 32–34.

Ekboir, J.M. (1999) The role of the public sector in the development and implementation of animal health policies. *Preventive Veterinary Medicine* 40(2), 101–115.

FAWC (2005) *Report on the Welfare Implications of Farm Assurance Schemes*. Farm Animal Welfare Council, London.

Foresight, UK (2011) *The Future of Food and Farming. Final Project Report*. The Government Office for Science, London.

Friggens, N., Kaya, C. and Roozen, A.W.M. (2017) GenTORE: genomic management tools to optimise resilience and efficiency. *Interbull Bulletin* 51, 91. Available at: http://www.gentore.eu/uploads/1/0/7/4/107437499/interbull_bulletin_no_51_tallinn_1.pdf (accessed 17 June 2020).

Garnett, T. and Godfray, C. (2012) Sustainable intensification in agriculture. navigating a course through competing food system priorities. *Food Climate Research Network and the Oxford Martin Programme on the Future of Food*, University of Oxford, UK, p. 51.

Gerber, P.J., Hristov, A.N., Henderson, B., Makkar, H., Oh, J. *et al.* (2013) Technical options for the mitigation of direct methane and nitrous oxide emissions from livestock: a review. *Animal* 7, 220–234.

Green, P.E. and Srinivasan, V. (1990) Conjoint analysis in marketing: new developments with implications for research and practice. *Journal of Marketing* 54(4), 3–19. DOI: 10.1177/002224299005400402.

Gunn, G.J., Stott, A.W. and Humphry, R.W. (2004) Modelling and costing BVD outbreaks in beef herds. *Veterinary Journal* 167(2), 143–149.

He, Y., Huang, H., Li, D., Shi, C. and Wu, S.J. (2018) Quality and operations management in food supply chains: a literature review. *Journal of Food Quality* 2018(2), 1–14. DOI: 10.1155/2018/7279491.

Hogeveen, H., Huijps, K. and Lam, T.J.G.M. (2011) Economic aspects of mastitis: new developments. *New Zealand Veterinary Journal* 59(1), 16–23.

Janssen, S.J.C., Porter, C.H., Moore, A.D., Athanasiadis, I.N., Foster, I. *et al.* (2017) Towards a new generation of agricultural system data, models and knowledge products: information and communication technology. *Agricultural Systems* 155, 200–212.

Kamilaris, A., Kartakoullis, A. and Prenafeta-Boldú, F.X. (2017) A review on the practice of big data analysis in agriculture. *Computers and Electronics in Agriculture* 143, 23–37. DOI: 10.1016/j.compag.2017.09.037.

Keeling, L. and Veissier, I. (2005) Developing a monitoring system to assess welfare quality in cattle, pigs and chickens. *Science and Society Improving Animal Welfare. Welfare Quality® Conference Proceedings*, Brussels, pp. 46–50.

Kehlbacher, A., Bennett, R. and Balcombe, K. (2012) 'Measuring the consumer benefits of improving farm animal welfare to inform welfare labelling'. *Food Policy* 37, 627–633.

Kragt, M.E., Pannell, D.J., McVittie, A., Stott, A.W., Ahmadi, B.V. *et al.* (2016) Improving interdisciplinary collaboration in bio-economic modelling for agricultural systems. *Agricultural Systems* 143, 217–224.

Lambe, N., Kenyon, F., Wishart, H., McBean, D., Waterhouse, T. *et al.* (2015) Eid and other technological advances in small ruminant research. *International Animal Health Journal* 2(4), 64–67.

Langford, F.M. and Stott, A.W. (2012) Culled early or culled late: economic decisions and risks to welfare in dairy cows. *Animal welfare* 21(1), 41–55.

Llonch, P., Haskell, M.J., Dewhurst, R.J. and Turner, S.P. (2017) Current available strategies to mitigate greenhouse gas emissions in livestock systems: an animal welfare perspective. *Animal* 11(2), 274–284.

Lusk, J.L. and Norwood, F.B. (2011) Animal welfare economics. *Applied Economic Perspectives and Policy* 33(4), 463–483.

McBean, D., Nath, M., Lambe, N., Morgan-Davies, C. and Kenyon, F. (2016) 'Viability of the Happy Factor™ targeted selective treatment approach on several sheep farms in Scotland'. *Veterinary Parasitology* 218, 21–30.

McInerney, J.P. (1993) Animal Welfare: An Economic Perspective. *Paper delivered to the Annual Conference of the Agricultural Economics Society*, Oxford, UK.

McInerney, J. (1996) Old economics for new problems livestock disease: presidential address. *Journal of Agricultural Economics* 47(1–4), 295–314.

McInerney, J.P. (1994) *Animal welfare, economics and policy. Report on a study undertaken for the Farm & Animal Health Economics Division of Defra.* Available at: https://webarchive.nationalarchives.gov.uk/20110318142209/http://www.defra.gov.uk/evidence/economics/foodfarm/reports/documents/animalwelfare.pdf

McInerney, J.P., Howe, K.S. and Schepers, J.A. (1992) A framework for the economic analysis of disease in farm livestock. *Preventive Veterinary Medicine* 13(2), 137–154. DOI: 10.1016/0167-5877(92)90098-Z.

Moran, D., Macleod, M., Wall, E., Eory, V., McVittie, A. *et al.* (2011) Marginal abatement cost curves for UK agricultural greenhouse gas emissions. *Journal of Agricultural Economics* 62(1), 93–118.

National Audit Office (2002) *The 2001 outbreak of foot and mouth disease.* Report by the Comptroller and Auditor General HC 939 Session 2001-2002.

Oliver, A. (2015) Nudging, shoving, and budging: behavioural economic-informed policy. *Public Administration* 93(3), 700–714.

O'Neill, J.I.M. (2014) Antimicrobial resistance: tackling a crisis for the health and wealth of nations. *Rev. Antimicrob. Resist* 20, 1–16.

Parasuaman, A., Zeithmal, V. and Berry, L.L. (1988) Multiple-items scale for measuring customer perceptions of service quality. *Journal of Retailing* 64(1), 12–40.

Ritson, C. (1978) *Agricultural Economics. Principles and Policy*, 2nd edn. Granada Publishing, London.

Robbins, L. (1935) *An Essay on the Nature and Significance of Economic Science*, 2nd edn. Macmillan, London.

Santarossa, J.M., Stott, A.W., Woolliams, J.A., Brotherstone, S., Wall, E. *et al.* (2004) An economic evaluation of long-term sustainability in the dairy sector. *Animal Science* 79(2), 315–325.

Serpell, J. (2004) Factors influencing human attitudes to animals and their welfare. *Animal Welfare* 13, S145–S151.

Smith, A. (1776) *An Inquiry into the Nature and Causes of the Wealth of Nations*, Cannan (ed). in 2 vols. Published by Simon and Brown 2012.

Stott, A.W. and Gunn, G.J. (2008) Use of a benefit function to assess the relative investment potential of alternative farm animal disease prevention strategies. *Preventive Veterinary Medicine* 84, 179–193.

Stott, A.W. and Gunn, G.J. (2017) Insights for the assessment of the economic impact of endemic diseases: specific adaptation of economic frameworks using the case of bovine viral diarrhoea. *Revue Scientifique et Technique de l'OIE* 36(1), 227–236.

Soteriades, A.D., Stott, A.W., Moreau, S., Charroin, T., Blanchard, M. *et al.* (2016) The relationship of dairy farm eco-efficiency with Intensification and self-sufficiency. Evidence from the French dairy sector using life cycle analysis, Data envelopment analysis and partial least squares structural equation modelling. *PLoS ONE* 11(11).

Stott, A.W., Lloyd, J., Humphry, R.W. and Gunn, G.J. (2003) A linear programming approach to estimate the economic impact of bovine viral diarrhoea (BVD) at the whole-farm level in Scotland. *Preventive Veterinary Medicine* 59(1–2).

Stott, A.W., Milne, C.E., Goddard, P.J. and Waterhouse, A. (2005a) Projected effect of alternative management strategies on profit and animal welfare in extensive sheep production systems in Great Britain. *Livestock Production Science* 97(2–3), 161–171. DOI: 10.1016/j.livprodsci.2005.04.002.

Stott, A.W., Coffey, M.P. and Brotherstone, S. (2005b) 'Including lameness and mastitis in a profit index for dairy cattle'. *Animal Science* 80(1), 41–52.

Stott, A.W., Vosough Ahmadi, B., Dwyer, C.M., Kupiec, B., Morgan-Davies, C. *et al.* (2012) Interactions between profit and welfare on extensive sheep farms. *Animal Welfare* 21(Suppl. 1), 57–64.

Tallentire, C.W., Edwards, S.A., Van Limbergen, T. and Kyriazakis, I. (2019) The challenge of incorporating animal welfare in a social life cycle assessment model of European chicken production. *The International Journal of Life Cycle Assessment* 24(6), 1093–1104.

Taylor, B.W. (2016) *Introduction to Management Science*, 12th edn. Pearson Education Ltd, Harlow, UK. Available at: https://www.amazon.co.uk/Introduction-Management-Science-Global-Bernard/dp/1292092912/ref=sr_1_3?keywords=management+science&qid=1561496600&s=books&sr=1-3

Thaler, R.H. and Sunstein, C.R. (2008) *Nudge: Improving Decisions About Health, Wealth and Happiness*. Yale University Press, London.

Tisdell, C. (1995) Assessing the approach to cost-benefit analysis of controlling livestock diseases of McInerney and others. Research Papers and Reports in Animal Health Economics 3. University of Queensland.

Toma, L., Stott, A.W., Revoredo-Giha, C. and Kupiec-Teahan, B. (2012) Consumers and animal welfare. A comparison between European Union countries. *Appetite* 58(2), 597–607.

Wemelsfelder, F., Hunter, T.E.A., Mendl, M.T. and Lawrence, A.B. (2001) Assessing the "whole animal": A free choice profiling approach. *Animal Behaviour* 62(2), 209–220.

Yalcin, C., Stott, A.W., Gunn, J. and Logue, D.N. (1999) The economic impact of mastitis-control procedures used in Scottish dairy herds with high bulk-tank somatic-cell counts. *Preventive Veterinary Medicine* 41, 135–149.

Index

Page numbers in **bold** refer to figures and tables.

CABI – who we are and what we do

This book is published by **CABI**, an international not-for-profit organisation that improves people's lives worldwide by providing information and applying scientific expertise to solve problems in agriculture and the environment.

CABI is also a global publisher producing key scientific publications, including world renowned databases, as well as compendia, books, ebooks and full text electronic resources. We publish content in a wide range of subject areas including: agriculture and crop science / animal and veterinary sciences / ecology and conservation / environmental science / horticulture and plant sciences / human health, food science and nutrition / international development / leisure and tourism.

The profits from CABI's publishing activities enable us to work with farming communities around the world, supporting them as they battle with poor soil, invasive species and pests and diseases, to improve their livelihoods and help provide food for an ever growing population.

CABI is an international intergovernmental organisation, and we gratefully acknowledge the core financial support from our member countries (and lead agencies) including:

Discover more

To read more about CABI's work, please visit: **www.cabi.org**

Browse our books at: **www.cabi.org/bookshop**,
or explore our online products at: **www.cabi.org/publishing-products**

Interested in writing for CABI? Find our author guidelines here:
www.cabi.org/publishing-products/information-for-authors/